Machine Learning in Multimedia

This book explores the interdisciplinary nature of machine learning in multimedia, highlighting its intersections with fields such as computer vision, natural language processing, and audio signal processing.

Machine Learning in Multimedia: Unlocking the Power of Visual and Auditory Intelligence serves as a comprehensive guide to navigating this exciting terrain where artificial intelligence meets the rich tapestry of visual and auditory data. At its core, this book seeks to unravel the mysteries and unveil the potential of machine learning in the realm of multimedia. Whether it's enhancing user experiences in virtual environments, revolutionizing medical diagnostics, or shaping the future of entertainment, the impact of machine learning in multimedia is profound and far-reaching. The journey begins with a thorough exploration of the foundational principles of machine learning, providing readers with a solid understanding of algorithms, models, and techniques tailored specifically for multimedia data. Through clear explanations and illustrative examples, readers will gain insights into how machine learning algorithms can be trained to extract meaningful patterns and insights from diverse forms of multimedia content. Moving beyond theory, this book delves into practical implementations and real-world applications of machine learning in multimedia. Through a series of case studies and examples, readers will witness firsthand how machine learning algorithms are transforming industries and reshaping the way we interact with multimedia content. Whether it's improving image recognition accuracy in autonomous vehicles, enabling personalized recommendations in streaming platforms, or enhancing speech recognition systems for better accessibility, the possibilities are limitless.

This book will be helpful to computer science, data science, and artificial intelligence researchers, students, and professionals looking to unlock the full potential of visual and auditory intelligence through the power of machine learning.

Innovations in Multimedia, Virtual Reality and Augmentation

Series Editors:
Rashmi Agrawal
Professor, Manav Racna International Institute of Research and Studies
Lalit Mohan Goyal
J.C. Bose Univ. of Sci. and Tech.

Data Visualization and Storytelling with Tableau
Edited By Mamta Mittal, Nidhi Grover Raheja

Multimedia Data Processing and Computing
Edited By Suman Swarnkar, J P Patra, Tien Anh Tran, Bharat Bhushan, Santosh Biswas

Artificial Intelligence in Telemedicine: Processing of Biosignals and Medical images
Edited By S. N. Kumar, Sherin Zafar, Eduard Babulak, M. Afshar Alam, Farheen Siddiqui

Multimedia Computing Systems and Virtual Reality
Edited By Rajeev Tiwari, Neelam Duhan, Mamta Mittal, Abhineet Anand, Muhammad Attique Khan

Advanced Sensing in Image Processing and IoT
Edited By Rashmi Gupta, Arun Kumar Rana, Sachin Dhawan, Korhan Cengiz

Data Visualization and Storytelling with Tableau
Mamta Mittal and Nidhi Grover Raheja

Machine Learning in Multimedia: Unlocking the Power of Visual and Auditory Intelligence
Edited by Suman Kumar Swarnkar, Annu Sharma, J. Somasekar, and Bharat Bhushan

For more information about this series, please visit: www.routledge.com/Innovations-in-Multimedia-Virtual-Reality-and-Augmentation/book-series/IMVRA

Machine Learning in Multimedia

Unlocking the Power of Visual and Auditory Intelligence

Edited by
Suman Kumar Swarnkar, Annu Sharma,
J. Somasekar, and Bharat Bhushan

CRC Press
Taylor & Francis Group
Boca Raton London New York

CRC Press is an imprint of the
Taylor & Francis Group, an **informa** business

Designed cover image: www.shutterstock.com (stock photo ID: 2365813093)

First edition published 2025
by CRC Press
2385 NW Executive Center Drive, Suite 320, Boca Raton FL 33431

and by CRC Press
4 Park Square, Milton Park, Abingdon, Oxon, OX14 4RN

CRC Press is an imprint of Taylor & Francis Group, LLC

ISBN: 978-1-032-76148-0 (hbk)
ISBN: 978-1-032-76147-3 (pbk)
ISBN: 978-1-003-47728-0 (ebk)

DOI: 10.1201/9781003477280

Typeset in Times New Roman
by Apex CoVantage, LLC

Contents

Preface

Welcome to the exciting world of machine learning in multimedia! In this rapidly evolving field, the convergence of visual and auditory intelligence with cutting-edge machine learning techniques has unlocked a wealth of possibilities across various domains, from entertainment and art to healthcare and security. This book serves as a comprehensive guide to understanding the fundamental concepts, methodologies, and applications of machine learning in multimedia. Whether you're a seasoned researcher, a curious student, or a practitioner looking to leverage the power of AI, this book is designed to provide you with a deep dive into the intersection of multimedia and machine learning. Throughout the chapters, we explore the foundational principles of machine learning, delving into algorithms, models, and techniques tailored specifically to multimedia data. From image classification and object detection to speech recognition and audio processing, each topic is meticulously crafted to provide both theoretical insights and practical implementations.

Moreover, this book goes beyond mere technical discussions. We delve into real-world case studies and applications, showcasing how machine learning is revolutionizing industries and reshaping the way we interact with multimedia content. Whether it's enhancing user experience in virtual reality environments or improving medical diagnostics through advanced imaging techniques, the potential of machine learning in multimedia knows no bounds. We would like to extend our gratitude to the contributors, researchers, and practitioners who have dedicated their time and expertise to advancing this field. Their insights and contributions have enriched the content of this book and helped shape its comprehensive scope.

Lastly, we hope that this book serves as a valuable resource on your journey through the fascinating realms of machine learning in multimedia. May it inspire you to explore new frontiers, innovate with confidence, and unlock the full potential of visual and auditory intelligence in the digital age.

Editor Biographies

Suman Kumar Swanrkar received a PhD (CSE) in 2021 from Kalinga University, Nayaraipur, Chhattisgarh. He received an MTech (CSE) in 2015 from the Rajiv Gandhi Proudyogiki Vishwavidyalaya, Bhopal, India. He has 12+ years of experience in educational institutes as an assistant professor and is currently associated with Shri Shankaracharya Institute of Professional Management and Technology, Raipur, as an assistant professor of computer science and engineering. He has guided 10+ MTech scholars, some of them ongoing. He has published and been granted an Indian/Australian patent. He has authored and co-authored more than 50 journal articles, including WOS and Scopus papers, and presented research papers at ten international conferences. He has completed many FDP training webinars and workshops and the two-week comprehensive online patent information course. He is proficient in handling teaching and research as well as administrative activities. He has contributed massive amounts of literature in the fields of intelligent data analysis, nature-inspired computing, machine learning, and soft computing.

Annu Sharma is currently working as Associate Professor in the Department of Computer Applications at PES University, Bangalore. She received a PhD in Computer Science from the Department of Computer Science, Gurukul Kangri University, Haridwar, Uttrakhand and Master's degree in Computer Science and Applications from the Department of Computer Science and Applications, University of Jammu, J&K. She has 22 years of teaching experience at the Master's level, including working Executives and Industry. She has worked with RRCE, Bengaluru, Bangalore University, Bengaluru, IMT Faridabad, Haryana, Central University of Jammu, J&K, and Arya College Ludhiana, Punjab. She has authored and co-authored more than 30 journal articles including SCI and Scopus Indexed papers, International book Chapters and has presented research papers at leading International Conferences. She has delivered various guest lectures at different colleges and is BoS for many education Institutions. She has acted as Session chair for various IEEE, National and International conferences. Her Research interest includes Spectral Biometrics, Computer Vision, Image Processing, Bioinformatics, AI and Machine Learning, Blockchain and IoT.

J. Somasekar received a PhD in CSE from JNTUA, Andhra Pradesh, and an MTech from the National Institute of Technology Karnataka (NITK), Surathkal. He is currently working as a professor of CSE, JAIN (Deemed-to-be University), Bangalore, and a postdoctoral researcher at the University of South Florida, USA. As a resource person, he has delivered 195 technical talks for FDPs, workshops, and webinars in 13 states of the country. He got an all-India rank of 43 in the GATE exam. He has 16 years of experience in teaching and 6 years of experience in research. He has published more than 35 research articles in leading journals indexed in SCI and Scopus and conference proceedings and three international textbook chapters. He is guiding five CSE PhD research scholars. His research interests include image processing, data science, machine learning, big data analytics, and ML for cybersecurity.

Bharat Bhushan is an assistant professor in the Department of Computer Science and Engineering (CSE) at the School of Engineering and Technology, Sharda University, Greater Noida, India. He received his undergraduate degree (BTech in computer science and engineering) with distinction in 2012, his postgraduate degree (MTech in information security) with distinction in 2015, and his doctorate (PhD in computer science and engineering) in 2021 from Birla Institute of Technology, Mesra, India. From 2021 to 2023, Stanford University (USA) listed Dr. Bharat Bhushan in the top 2% of scientists. He earned numerous international certifications, such as CCNA, MCTS, MCITP, RHCE, and CCNP. He has published more than 150 research papers at various renowned international conferences and in SCI-indexed journals. He has contributed more than 50 chapters in various books and has edited 30 books from the most famed publishers. He is a series editor of two prestigious Scopus Indexed Book Series, Computational Methods for Industrial Applications (CMIA) and Future Generation Information Systems (FGIS), published by CRC Press, Taylor and Francis, USA. He has served as keynote speaker (resource person) in numerous reputed faculty development programs and international conferences held in different countries, including India, Iraq, Morocco, China, Belgium, and Bangladesh. He has served as a reviewer/editorial board member for several reputed international journals. In the past, he worked as an assistant professor at HMR Institute of Technology and Management, New Delhi, and network engineer at HCL Infosystems Ltd., Noida.

Contributors

Bhushan, Bharat
Department of Computer Science and Engineering (CSE)
School of Engineering and Technology
Sharda University
Greater Noida, India

Bansal, Poonam
Artificial Intelligence and Data Sciences
Indira Gandhi Delhi Technical University for Women
Delhi, India

Bansal, Tanisha
Artificial Intelligence and Data Sciences
Indira Gandhi Delhi Technical University for Women
Delhi, India

Dixit, Rohit R
Siemens Healthineers
Boston, Massachusetts

Gulati, Neha
University Business School Panjab University
Chandigarh, India

Gulati, Sunidhi
Artificial Intelligence and Data Science
IGDTUW
New Delhi, India

J. Somasekar
JAIN (Deemed-to-be University)
Bangalore, India

Malik, Kiran
Artificial Intelligence and Data Sciences
Indira Gandhi Delhi Technical University for Women
Delhi, India

Negi, Nishtha
Indira Gandhi Delhi Technical University for Women
Kashmere Gate, Delhi, India

Preeti
Department of Computer Science and Engineering
Maharishi Dayanand University
Rohtak, India

Punia, Khushi
Computer Science Engineering – Artificial Intelligence
Indira Gandhi Delhi Technical University For Women
Delhi, India

Rahman, Habib Ur
Department of Computer Science and Engineering
Kanpur Institute of Technology
Kanpur, India

Rana, Chhavi
Department of Computer Science and Engineering
Maharishi Dayanand University
Rohtak, India

Rathore, Yogesh Kumar
Shri Shankaracharya Institute of Professional Management and Technology
Raipur Chhattisgarh, India

Reddy, SRN
Indira Gandhi Delhi Technical University for Women
Kashmere Gate, Delhi, India

Rustagi, Tanvi
K. R Mangalam University
Gurugram, Haryana, India

Sharma, Annu
Department of Computer Applications
PES University
Bengaluru, India

Sharan, Shambhu
Centre of Excellence – Artificial Intelligence
Indira Gandhi Delhi Technical University for Women
Delhi, India

Sharma, Harshita
IGDTUW
New Delhi, India

Singh, Yudhvir
Computer Science and Engineering
M.D. University
Rohtak, Haryana, India

Srivastava, Mayank
Computer Science
The NorthCap University
India

Sujata
Computer Science
The NorthCap University
India

Suman
Department of Computer Science and Engineering
M.D. University
Rohtak, Haryana, India

Suman, Ruchika
IITM Janakpuri
New Delhi, India

Swarnkar, Suman Kumar
Shri Shankaracharya Institute of Professional Management and Technology
Raipur Chhattisgarh, India

Verma, Richa
IGDTUW
New Delhi, India

Vijarania, Meenu
Centre of Excellence – CSE
K.R. Mangalam University
Gurugram, Haryana, India

Yadav, Jyotsna
Computer Science and Engineering
Kanpur Institute of Technology
Kanpur, India

Yajur
DPS Sushant Lok
India

1 Machine Learning Techniques for Accurate Prediction and Detection of Chronic Diseases

Suman, Yudhvir Singh, Neha Gulati

1.1 INTRODUCTION

As a consequence of modern society's norms and practices, humans are more susceptible to several diseases. Avoiding the worst effects of these illnesses requires early identification and prognosis. Manual diagnosis presents significant challenges for practitioners. This chapter aims to identify and predict who will be affected by the most common chronic diseases. To guarantee that this classification accurately identifies people with chronic conditions, a cutting-edge machine learning technique will be used. Disease prognosis is another complex endeavour. Because of this, data mining is crucial for disease forecasting, as shown in Figure 1.1.

Using machine learning techniques, the suggested system delivers a thorough illness prediction based on the patient's symptoms. Worldwide, the prevalence of chronic diseases is a big challenge for healthcare systems [1]. According to the medical assessment, chronic diseases are mostly to blame for the increasing human death rate. The expense of care for these conditions often consumes over 70% of a patient's income [2]. Figure 1.1 presents some chronic diseases. It is crucial to minimize the patient's exposure to potentially fatal risks. Health data is becoming easier to collect as a result of technological advancements. Healthcare data includes patient demographics, analytical medical reports, and sickness histories. Different geographies and different types of environments may result in different kinds of illnesses [3]. Therefore, the patient's living conditions and environmental factors should be included alongside the illness information. Figure 1.2 presents examples of chronic kidney disease (CKD), for which SVM and KNN algorithms, a decision tree, and random forest classifiers are considered [4].

1.1.1 Chronic Disease

Chronic illnesses are those that endure longer than three months, as defined by the US National Centre for Health Statistics [5]. No drugs or vaccinations exist to combat these illnesses. Tobacco use, poor nutrition, and insufficient exercise are the leading

DOI: 10.1201/9781003477280-1

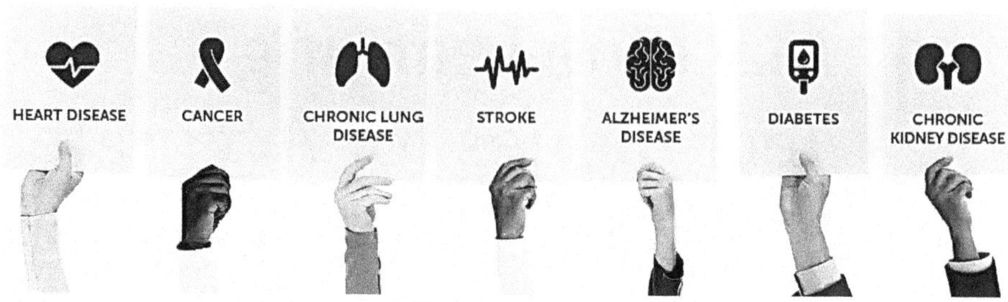

FIGURE 1.1 Chronic disease.

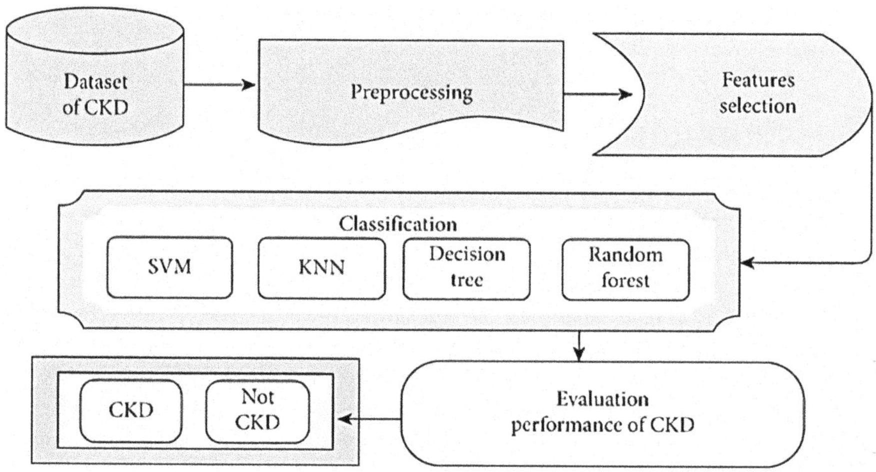

FIGURE 1.2 Chronic kidney disease detection using ML.

causes of chronic illness. In addition, ageing is a prevalent factor in the development of this condition. Heart disease, cancer, arthritis, diabetes, obesity, epilepsy, seizures, and dental issues are some chronic illnesses. Heart disease and stroke are two examples of cardiovascular disease that can prove fatal. Tobacco usage, eating poorly, and not getting enough exercise all contribute to these illnesses [6]. By adjusting these behaviours, patients may lessen their influence on managing and avoiding cardiovascular disease. Cancers like colon and breast cancer are considered major causes of death in the world. Only with precaution, early diagnosis, and adequate medical care can they be contained. Reducing exposure to carcinogenic environments and lifestyle choices is one way to lower cancer risk [7].

Inflammation of the joints, the source of arthritis pain and stiffness, worsens with age. Low-cost strategies for alleviating arthritic symptoms are at hand but are underutilized [8]. Regular, moderate exercise has been shown to lessen the symptoms of arthritis. Diabetes is a costly and life-threatening condition. The effects of diabetes may be mitigated with self-care and early diagnosis. Seven million Americans aged 65 and above have diabetes, with the majority having type 2. The prevalence of obesity among persons of all ages has increased since 1980. One's chance of developing hypertension (BP), cardiovascular disease (CVD), diabetes, and rheumatoid arthritis increase with weight gain [9]. Some forms of cancer are also linked to obesity. The expense of treating epilepsy and seizures is quite considerable. This illness affects people of all ages, but particularly the young and the old. When it comes to the well-being of the elderly, oral health issues are given a lot of attention. A person's ability to talk, chew, swallow, and keep to a healthy diet are all compromised, making this a very significant problem [10].

1.1.2 Role of ML in Chronic Disease Prediction

From cutting-edge gadgets to medical applications (such as illness detection and patient safety), machine learning (ML) is ubiquitous. ML is gaining popularity in several fields, and medical diagnosis is only one of them. Many researchers and medical professionals have shown the potential of using machine learning in illness diagnosis. Diagnosis using conventional methods is labor intensive, time consuming, and expensive. Unlike conventional methods of diagnosis, which are limited by the practitioner's skill level, ML-based systems are not susceptible to fatigue [11]. Therefore, a strategy for sickness diagnosis in areas with an inadequate patient population may be developed. These systems are built with the use of X-ray and MRI images, together with tabular data on patients' diseases, ages, and sex [12]. ML is a branch of artificial intelligence that takes data as input. It is sometimes difficult for people to achieve the results that may be obtained by using specified mathematical functions. Malignant cells in a microscopic picture, for instance, maybe more easily identified using ML than they would be by just visually inspecting the image. Figure 1.2 shows chronic kidney disease detection using ML [13, 14].

1.1.3 Motivation of Research

Integrating IT has led to faster growth in the healthcare industry in recent years. Similar to how smartphones have simplified daily living, the goal of incorporating IT into healthcare is to lower costs without sacrificing quality of treatment or patient comfort. Intelligent healthcare might make this a reality by providing several benefits to both patients and medical professionals. Patients with chronic illnesses in a certain area show little variation between sexes, according to annual studies conducted on the subject, and in 2014, a record number of people were admitted to hospitals for care of these conditions. When compared to only utilizing structured data, outcomes based on a combination of the two are far more reliable. Rare diseases are often difficult to diagnose. Therefore, self-reported behavioural data may be used to separate those with rare diseases from those with more prevalent chronic ailments. It is hoped that using ML algorithms and questionnaires will allow for the reliable diagnosis of rare diseases.

1.2 LITERATURE REVIEW

This section details the complementary efforts made in creating the suggested model for chronic illness prediction. The following discussions were conducted after reviewing the current literature to assist in the development of the proposed system.

In [15], we reviewed the methods of ML for assessing the risk of chronic kidney disease. Our goal is to develop effective CKD prediction tools using ML methods. To address the uneven distribution of cases between the two classes, we first use class balancing; next, we rank and analyze features; and lastly, we train and assess several ML models using different measures of success. According to the findings, RotF was the most effective of the analyzed models, with 100% AUC, recall, precision, F measure, and accuracy

Reference [16] introduced the LSTMNN. They can identify artifacts in long-term recordings of local field potentials. This study compares the classification performance and computational time of a feed-forward NN with that of a RNN-based ML approach: specifically a configuration of RNN known as LSTM used in two configurations to detect patterns. Findings demonstrate that the LSTM model is capable of providing accuracy of 87.1%.

The internet of medical things adaptive hybridized DCNN for prediction of chronic kidney disease. In this study, we compare many deep learning approaches and propose an AHDCNN for accurate and efficient early diagnosis of kidney illness. The effectiveness of classification methods is context dependent. A model approach was built utilizing CNN to minimize feature dimension and improve the classification system's accuracy.

Reference [17] reviewed ML prediction models for the prognosis of chronic illness. Our document search was done from the libraries of PubMed and CINAHL, and it comprises 453 publications published between 2015 and 2019. In the end, 22 papers were chosen to precisely illustrate all modeling methodologies, including their benefits and drawbacks, in order to explain CD diagnostic and use models of distinct illnesses. Given that every strategy has its pros and cons, our results show that there are no universally accepted ways to decide which approach is preferable in actual clinical practice.

Reference [18] introduced the use of soft clusters and focused on improving ML algorithms in detection of chronic diseases. The author considered data on chronic diseases from Kaggle. Simulation results show that mechanism was effective in detection of chronic illness.

Reference [19] presented work on AI for diabetes. They provided an in-depth analysis of how DL has been used in the study of diabetes. They did a thorough literature search and found that this method was mostly used in three settings: diabetes diagnosis, glucose control, and the detection of complications from diabetes. After doing the search, they narrowed down the results to 40 publications and have provided a summary of the most important details, including the learning mode, the development process, the primary outcomes, and the baseline techniques for performance measurement. Among the reviewed papers, it stands out that several deep learning algorithms and frameworks have outperformed traditional ML approaches to attain state-of-the-art performance on numerous problems involving diabetes.

Reference [20] reviewed the prediction of several diseases using LSTM neural networks. They demonstrated that our suggested strategy outperforms a number of baseline and DL approaches to illness prognosis forecasting. They next compared the proposed attention-based mechanism's performance to that of previously studied attention-based mechanisms and investigated the effects of various time interval options for the time-aware mechanism. Our findings have applications in helping doctors with their diagnosis and, more generally, in enhancing the standard of care provided by medical facilities.

Reference [21] contains research on DL models that are still interpretable to detect patients with persistent cough using EHR data. The prevalence of CC among adults is estimated to be 10%. Asthma, bronchitis, and gastro-oesophageal reflux disease were just a few of the many diseases linked to a persistent cough. Because there isn't an ICD code for persistent cough, it's difficult to get that information from EHRs. There was an immediate demand for computational approaches using EHR data to detect instances of persistent cough for use in clinical and research settings. In order to forecast persistent coughs, this study will look at the data representations and DL techniques used for this task

Reference [22] provided work on cardiovascular disease used LSTM deep learning models. The enhanced LSTM model was the primary emphasis of this article as it related to the prediction of cardiovascular disease. This research offered a novel model based on the classic LSTM by enhancing the forgetting gate input inside the LSTM itself. To get over the difficulty in prediction that comes from the irregular time interval, we first smoothed it out to produce the time parameter vector and then fed it into the forgetting gate.

Reference [23] included automated kidney disease detection using a bidirectional LSTM-based correlative NN. A CKD sensing module is used for training and testing the suggested model. Because they can find the most useful characteristics in the data, DL algorithms can boost detection precision. The suggested technique averaged a 98.08% success rate on the test dataset.

Reference [24] introduced the analysis of current uses of AI in epidemic disease detection with a focus on COVID-19 and other pathogens. In this article, they explain use of NN and ML in the diagnosis and detection. The author focused on detection of COVID-19 and other diseases, as well as the nations involved in the development of these methods. There were examples of how RNN, DNN, and multi-layer CNN may be used to identify and forecast diseases like COVID-19. Datasets for COVID-19 are described, along with the research and analytical capabilities they provide. Real-time COVID-19 data is used to demonstrate applications of AI and NN.

Reference [25] suggested work on predicting cardiac issues using DL in a smart healthcare monitoring system based on IoT and the cloud. Data from IoT devices is gathered by the proposed system, and predictive analytics are used on the patient's electronic healthcare data on the cloud. The Bi-LSTM-based smart healthcare system is proposed for monitoring. This system is also focused on accurate prediction for heart disease risk and exhibits superior accuracy (98.86%), precision (98.9%), sensitivity (98.8%), specificity (98.89%), and F-measure (98.86%) and is the state of the art in this field.

Reference [26] looked at disease diagnosis using AI, synthesis framework, and research goals for the future. They surveyed extensively, covering everything from the medical imaging datasets people were using to how they were extracting and classifying features for predictions. Following the recommended reporting elements for systematic reviews and meta-analyses, we used the Web of Science and Psychology Information to choose articles on the topic of AI-based early illness prediction. The diagnostic articles were evaluated using a variety of quality criteria.

Reference [13] focused on ML for diagnosis of chronic renal disease. In this investigation, authors used classification mechanisms for stage prediction. RF, SVM, and DT were the prediction models used. For feature selection, they used analysis of variance and iterative feature removal with cross-validation. Experiments showed that recurrent feature elimination with cross-validation--based RF outperformed support vector machines and decision trees.

Reference [27] reviewed the methodology development for disease prediction using R look-ahead LSTM. The predictive power of a model for cardiovascular illness was improved with the help of R look-ahead LSTM. After optimizing the LSTM network model, a cardiovascular disease prediction model is proposed.

Reference [28] introduced analytics for the future of healthcare based on ML and DNN. The purpose of this article was to analyze current ML and DL methods used in healthcare prediction and to highlight the challenges that arise when attempting to use these methods in this setting. The article provides a thorough overview of field and discusses the current difficulties associated with healthcare prediction. AI has been found to have a significant influence on the medical diagnosis process.

Reference [29] includes research on chronic disease detection and prognosis using ML. The suggested system delivers a thorough illness prognosis. This general illness prediction takes into consideration the person's lifestyle, information about any recent medical issues, and the results of a symptom collection undertaken in preparation for the dataset. This research concludes with a comparison of the proposed method to other popular ML techniques, including Nave Bayes, DT, and logistic regression.

Reference [30] introduced disease diagnosis using ML. Using information from Scopus and WOS, a bibliometric analysis of publication was performed. To identify the most prolific authors and countries, a bibliometric analysis of 1,216 publications was conducted. The most up-to-date developments and strategies in MLBDD were then summarized, with consideration given to algorithms, kinds of illness, and assessment metrics. This study concludes with a discussion of major findings and a look forward at potential developments in the field of MLBDD

Reference [31] focused on multi-task intrapersonal learning for the early diagnosis of chronic illness. Research focused on improving the model's performance by collaborative prediction of the state of linked chronic conditions. In order to make an accurate forecast, they took into account variables specific to each chronic condition as well as the temporal connection of the time-series data. The study consisted of three phases: data preprocessing and feature selection using BRITS and the LASSO, CNN-LSTM for single-task models, and multi-task learning.

1.3 PROBLEM STATEMENT

The issue with conventional research is that research related to chronic diseases is limited and has limited accuracy and performance. Conventional model are supposed to be upgraded in order to get better accuracy with reduced error rates. Moreover, there is a need to enhance the performance of DL models during training and testing. Thus, there is a need to integrate LSTM into CNN in order to propose CNN-LSTM models to assure reliability.

1.4 COMPARATIVE ANALYSIS OF CONVENTIONAL MECHANISMS

Table 1.1 presents a literature survey. The research is a comparison of the analysis of conventional mechanisms, including methodology and limitations.

TABLE 1.1
Literature Survey

Sno.	Author	Year	Title	Methodology	Limitation
1	Fabietti	2020	LSTM, Neural Network for Artefact Detection in Continuously Recorded Local Field Potentials	LSTM, chronically	This is not long-lasting work.
2	Chen	2020	Prediction of Chronic Kidney Disease by Making Use of Adaptive Hybridized Deep CNN	AHDCNN	There is lack of security and salability.
3	Battineni	2020	Predictive Models for Chronic illness Detection Using Machine Learning	CNN and KNN	There is a lack of technical work.
4	Aldhyani	2020	Using Soft Clustering to Improve Chronic Disease Diagnosis with ML Algorithms	ML	Lack of flexibility.
5	Zhu	2021	To Review on Diabetes and Deep Learning	Deep learning	Need to do more work on accuracy.
6	Men	2021	Predicting the Onset of Many Diseases Using LSTM	LSTM, RNN	Need to enhance the scope of work.
7	Luo	2021	Using EHR Data to Identify Patients with Persistent Cough Using Interpretable Deep Learning Models	Deep learning	Need to consider optimization techniques.
8	Divya Vani	2021	Predicting Cardiovascular Disease Using LSTM	LSTM, DL	Did not considered real- life solutions.
9	Bhaskar	2021	A Bidirectional LSTM-Based Correlational NN for Automated Kidney Disease Detection	LSTM	Performance of this research is very low.
10	Segall	2021	Artificial Intelligence in the Early Detection and Analysis of COVID-19: A Review	AI	Lack of technical work.
11	Nancy	2021	Predicting Cardiovascular Disease with an IoT and Cloud-Based Health Monitoring System Using DL	Deep learning	Lack of security and accuracy.
12	Manjula	2022	Using Deep Learning for Detection of Multiple Diseases	Deep learning	There is a lack of performance.
13	Kumar	2022	Disease Diagnosis Using Artificial Intelligence: A Literature Review, Synthesis Framework, and Research Roadmap	AI	There is less technical work.
14	Dritsas	2022	Applications of ML to Prediction of Chronic Kidney Disease Risk	ML	Research is limited to traffic flow.
15	Debal	2022	ML Methods for Predicting Chronic Kidney Disease	ML	There is a lack of performance.
16	Badawy	2022	Application of ML and DL to Predictive Analytics in Healthcare	ML, DL	Lack of efficiency.
17	Alanazi	2022	Chronic Diseases: Using an ML Approach for Detection and Prediction	ML, chronic diseases	There is a lack of performance.
18	Ahsan	2022	Artificial Intelligence–Based Illness Probable Cause: A Comprehensive Review	ML, disease diagnosis	Needs to focus on influencing factors that are affecting performance.
19	Kim	2023	Predicting Chronic Diseases Using Intra-Person Multi-Task Learning	Data preprocessing using BRITS and LASSO; CNN-LSTM models	Scope of work is limited.

1.5 NEED FOR RESEARCH

The suggested method aims to detect and forecast an individual's risk of developing a chronic illness by using a machine learning strategy. Structured data, unstructured data, and lifestyle choices are included in the dataset. The missing values in this data have already been preprocessed. They are then recreated to boost the model's quality and accuracy of prediction. Machine learning algorithms are employed for prediction. Table 1.2 presents features considered in conventional research. There is a need to propose an advanced approach to chronic disease detection using machine learning. Comparisons of features in existing researches are expressed in Table 1.2.

TABLE 1.2
Comparisons of FeaturesT

Citation	Chronic Diseases	Machine Learning	LSTM	AI	Deep Learning
[1]	✓	✗	✗	✗	✗
[2]		✓	✗	✗	✗
[3]	✓	✗	✗	✗	✗
[4]	✓	✗	✗	✗	✗
[5]	✓	✓	✗	✗	✗
[6]	✗	✓	✗	✓	✗
[7]	✓	✗	✗	✗	✓
[8]	✗	✓	✗	✗	✗
[9]	✓	✓	✗	✗	✗
[10]	✓	✗	✗	✗	✓
[11]	✓	✗	✗	✗	✗
[12]	✓	✗	✗	✗	✗

1.6 CONVENTIONAL CLASSIFIERS APPLIED FOR IDENTIFICATION AND PREDICATION OF CHRONIC DISEASE

By examining data samples and drawing main inferences using mathematical and statistical approaches, machine learning enables computers to learn without being explicitly programmed [32]. With the use of games and pattern-recognition algorithms, Arthur Samuel first presented machine learning in 1959. Depending on the task at hand, machine learning is built on the utilization of data for prediction or decision-making [33]. Many traditionally time-consuming activities are now readily and rapidly accomplished thanks to machine learning technology [34]. As both computational power and data storage capacity increase exponentially, it is simpler to train data-driven ML models to anticipate occurrences with near-perfect accuracy. Many researches provide an abundance of ML techniques. Research categorizes these methods into three broad categories: supervised, unsupervised, and semi-supervised algorithms [35], as shown in Figure 1.3.

1.6.1 SUPERVISED MACHINE LEARNING

In supervised learning, also known as supervised machine learning, labelled datasets are used to train computers how to efficiently categorize data or predict outcomes. Regardless of the data that is being fed into it, the model will continue to adjust its weights until it has been fitted appropriately. This will occur until the model is complete. This step is conducted as part of the process of cross-validation to guarantee that the model does not suffer from either overfitting or underfitting (Zhang et al. 2019). The goal of this phase is to ensure that the model accurately represents the data. Businesses have the ability to solve a broad variety of real-world difficulties at scale utilizing supervised learning. One example of this would be the separation of spam emails into a separate folder from other incoming communications. According to Punia and Mittal (2014), examples of techniques used in supervised learning

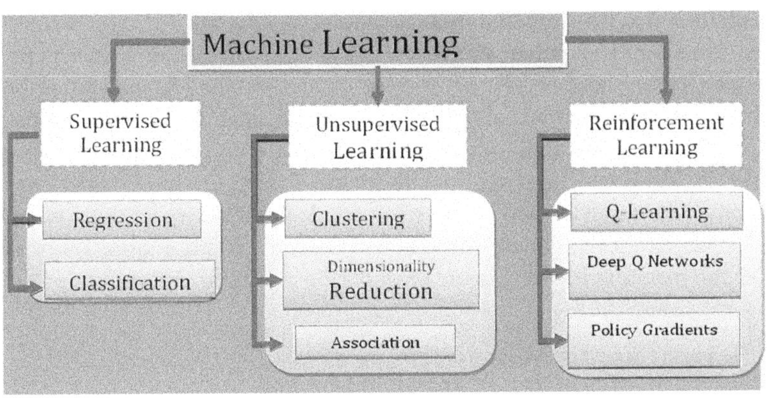

FIGURE 1.3 Types of machine learning.

include neural networks, Naive Bayes, linear regression, logistic regression, random forest, and support vector machine (SVM). These are only few of the possibilities.

1.6.2 Unsupervised Machine Learning

Unsupervised learning, which is also known as unsupervised machine learning, is a type of learning that does not include human supervision and makes use of machine learning algorithms. This kind of learning is utilized to assess and cluster information that has not been labelled. These algorithms reveal previously undiscovered patterns or data groupings without requiring any participation from a human researcher in any way. Due to its ability to unearth parallels and differences in the content that is being studied, this method is very useful for doing exploratory data analysis, cross-selling tactics, customer segmentation, picture identification, and pattern recognition, to name a few of its many potential uses. According to Bradley et al. (2019), it is also possible to utilize it to reduce the number of features included in a model by employing a technique known as dimensionality reduction. Principal component analysis (PCA) and singular value decomposition (SVD) are two approaches that are frequently used for this purpose. According to Mittal, Kumar, and Vijayalakshm (2014), unsupervised learning may also make use of other kinds of algorithms, such as neural networks, k-means clustering, and probabilistic clustering techniques.

1.6.3 Semi-Supervised Machine Learning

The benefits of learning in a supervised environment as well as those of learning in an uncontrolled environment can be balanced out by learning in a semi-supervised environment. While it is being trained, it makes use of a smaller dataset that has been labelled to drive classification and feature extraction from a larger dataset that has not been labelled. This is done in comparison to the larger dataset. If an algorithm designed for supervised learning does not have access to a significant amount of data that has been tagged, semi-supervised learning may be utilized to circumvent this problem. According to Hong et al. (2019), it is also helpful in cases in which it would be prohibitively expensive to label sufficient amounts of data.

Nonetheless, it is possible that there are various classes of ML algorithms that are grounded in distinct learning modalities (Tang et al., 2019). Most modern DL models are built on CNNs; these can be used alone or in combination with other DL models, such as generative models, deep belief networks, and the Boltzmann machine to extract information from input (such as images, numbers, and classes) (Dritsas and Trigka 2022). There are three distinct varieties of deep learning: supervised, semi-supervised, and unsupervised. The three most well-known deep learning architectures are DNNs, RL, and RNN (Javid, Alsaedi, and Ghazali, 2020). Figure 1.4 presents the model for the machine learning cycle considering selection algorithms.

1.6.4 Support Vector Machine

Finding the optimal line (or decision boundary) that divides the n-dimensional space into classes is the objective of the support vector machine (SVM) approach. This will

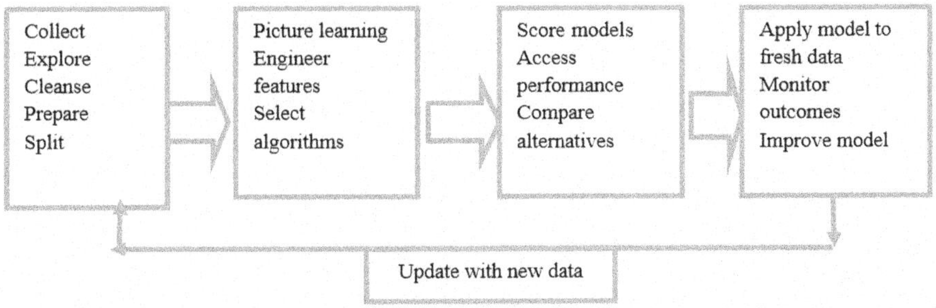

FIGURE 1.4 Model for machine learning cycle.

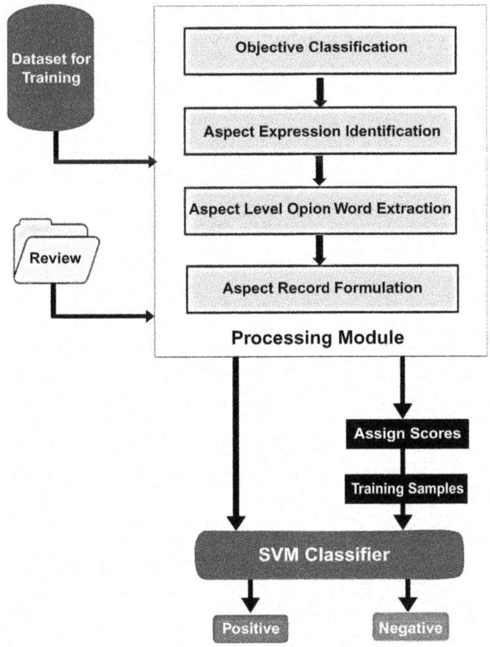

FIGURE 1.5 Model of SVM.

make the classification of data points in the future much simpler. When it comes to making a choice, a hyperplane may be used to define the optimal range of options. When the data can be broken down linearly or non-linearly, the results are far more accurate. When the data can be partitioned into two classes along a straight line, SVMs will provide a separating hyperplane as their output [36]. This hyperplane will optimize the margin of separation that exists between the different classes as shown in Figure 1.5.

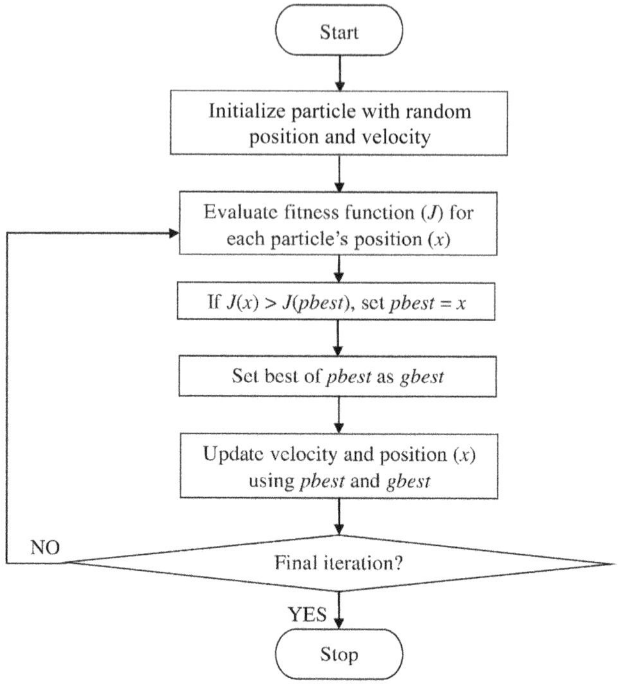

FIGURE 1.6 Flowchart of PSO algorithm.

Due to the fact that it is an SML technique, SVM may be used for both classification and regression. It can make sense of your data and select an appropriate cutoff for the range of outcomes because it applies the kernel technique to your data. In order to optimize SVM, particle swarm optimization is applied [37]. A flowchart showing the optimized SVM with fitness function to find particle position is shown in Figure 1.6.

Objective function used during optimization

```
function [o] = ObjectiveFunction (x)
% o = sum (abs(x)) + prod(abs(x));
o = (mean(abs(x)) + max (abs(x)))/2;
end
```

1.6.5 Naïve Bayes

NB classifiers are a family of straightforward "probabilistic classifiers" in the field of statistics, as shown in Figure 1.7. They are based on Bayes's theorem and operate on the premise that individual features are independent of one another.

When combined with KDE, the simple Bayesian network models shown in Figure 1.7 have the potential to attain a high degree of accuracy. Classification techniques based on Bayes's theorem include the Naive Bayes classifier (Punia and

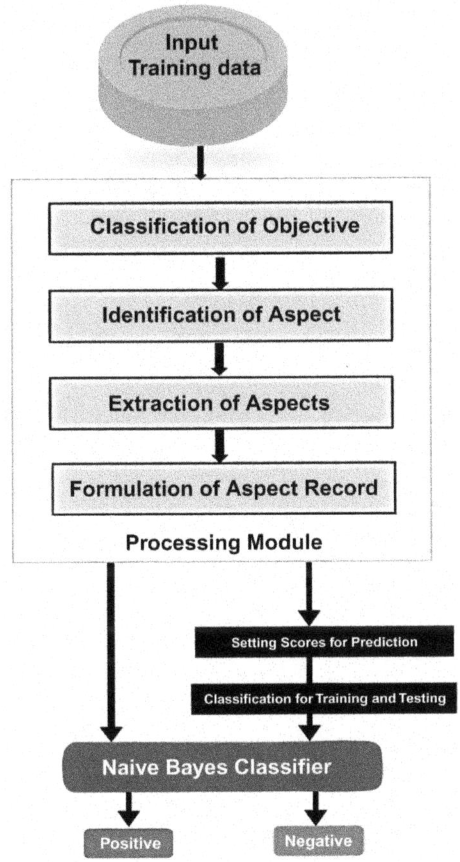

FIGURE 1.7 Naive Bayes model.

Mittal, 2014). There is not just one algorithm here; rather, there is a family of systems that all share the same working assumption. This assumption is that no two sets of characteristics that are being categorized are in any way related to one another.

ALGORITHM TO MAKE PREDICTION USING NAÏVE BAYES

Input:

Training dataset T,
F= (fi, f2. f3, . . . fa) in testing dataset.
// value of the predictor variable

Output:

A class of testing dataset

Step:

1. Obtain the dataset that will be used for training.
2. Determine mean and SD of predictor variables for each class.
3. Repeat.
   Calculate probability of f by applying Gauss density equation to each class. Continue doing this until probability of all predictor variables (fi, f2, f, etc.) has been determined.
4. Determine the probability of occurrence for each group.
5. Achieve the highest possible probability.

1.6.6 LSTM

LSTM is useful for time series forecasting models because it enables them to extrapolate values from a given sequence of data. If demand forecasting can be

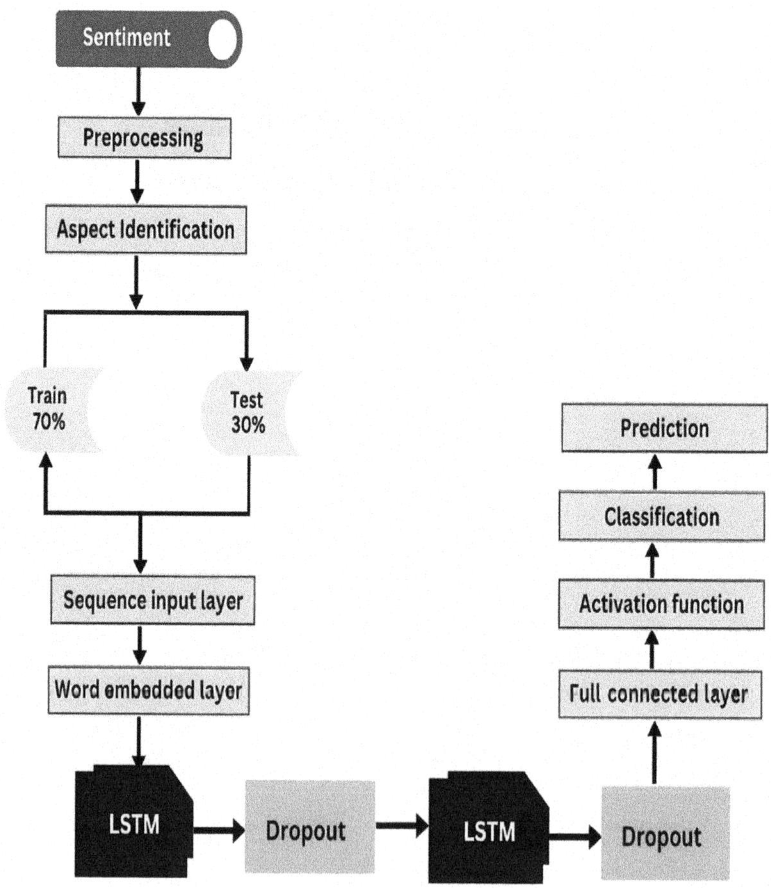

FIGURE 1.8 LSTM model.

more accurately done, it's possible that corporate leaders will be able to make more informed choices, as shown in Figure 1.8.

When training a network on lengthy sequences of words or numbers, you may run into issues due to a phenomenon known as the vanishing gradient, which may be circumvented with the help of LSTM networks. Results obtained with LSTM are almost always superior to those obtained using SVM because it is superior in terms of the amount of information it can either keep or throw out. The SVM model and the LSTM model both gain a great deal from the use of moving averages; nevertheless, as can be seen in Figure 1.6, the SVM model performs substantially better on a combined dataset than the LSTM model does on the standard base dataset. The use of PSO for LSTM model optimization is shown in Figure 1.9.

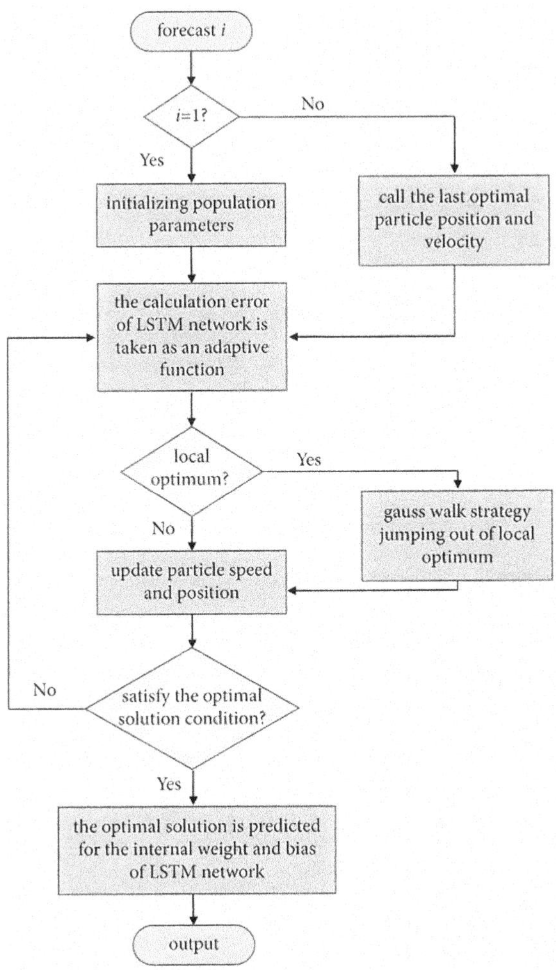

FIGURE 1.9 Optimization of LSTM model using PSO.

1.7 RESEARCH METHODOLOGY

In this research, conventional ML and DL models are considered for detection and classification of chronic disease. Issues with conventional work include lack of accuracy, error rate, and performance. Present research is simulating classifiers such as SVM, Naïve Bayes, and LSTM on datasets of chronic disease. Finally, the accuracy and performance of all three models are compared. Figure 1.10 presents the process flow of the proposed work.

1.8 RESULT AND DISCUSSION

In a dataset including chronic diseases, the current study involves modeling several classifiers, including SVM, naïve Bayes, and LSTM. In the last step, we evaluate and contrast the accuracy, error rate, and performance of all three models.

1.8.1 Accuracy

Simulation uses a seven-fold cross-validation because there is a probability increment in accuracy. Table 1.3 presents the comparative analysis of accuracy for SVM, Naïve Bayes, and LSTM models for all these seven-fold cross-validations.

Figure 1.11 presents graphical diagram considering Table 1.3.

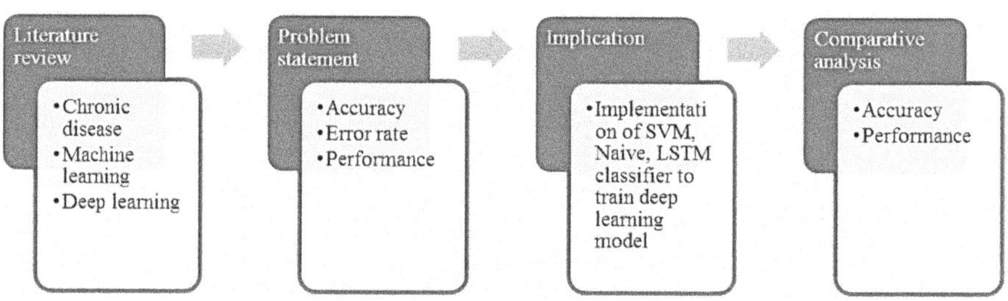

FIGURE 1.10 Process flow of proposed work.

TABLE 1.3
Comparison of Accuracy

Split	SVM Model	Naïve Bayes Model	LSTM Model
1	89.17%	91.69%	93.21%
2	89.13%	91.84%	93.33%
3	89.48%	91.71%	93.72%
4	89.92%	91.42%	93.62%
5	89.74%	91.07%	93.93%
6	89.44%	91.86%	93.56%
7	89.14%	91.88%	93.50%

1.8.2 Error Rate

Table 1.4 presents a comparative analysis of the error rates.

Figure 1.12 presents a graphical diagram considering Table 1.4.

1.8.3 Time Taken

Machine specification: Core I5, Ram 16 Gb, Storage space 1 TB

Table 1.5 presents the comparative analysis of time taken.

Figure 1.13 presents a graphical diagram considering Table 1.5.

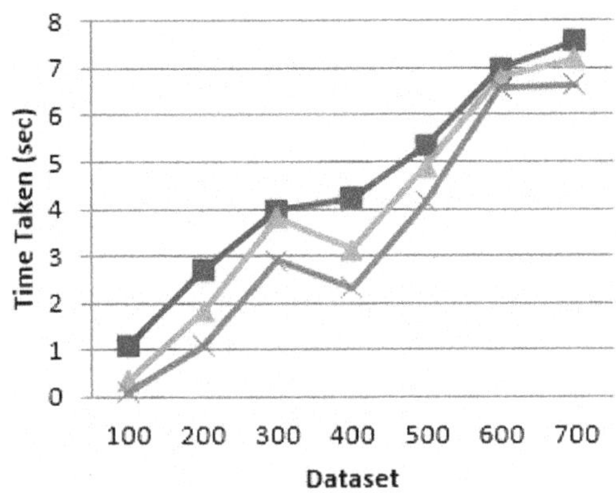

FIGURE 1.11 Comparison of accuracy.

TABLE 1.4
Comparison of Error Rates

Split	SVM Model	Naïve Bayes Model	LSTM Model
1	10.83%	8.31%	6.79%
2	10.87%	8.16%	6.67%
3	10.52%	8.29%	6.28%
4	10.08%	8.58%	6.38%
5	10.26%	8.93%	6.07%
6	10.56%	8.14%	6.44%
7	10.86%	8.12%	6.50%

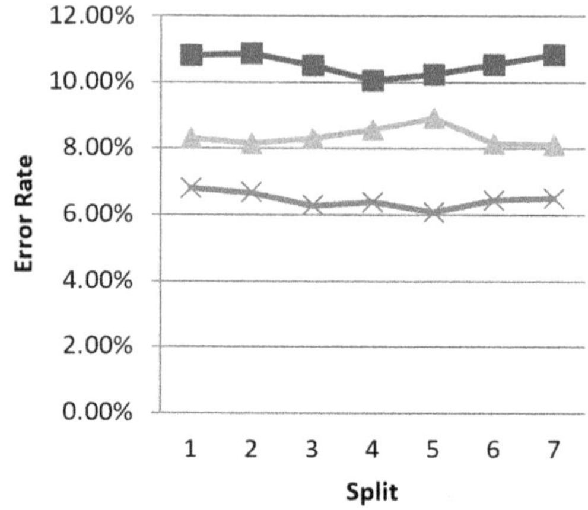

FIGURE 1.12 Comparison of error rates.

TABLE 1.5
Comparison of Time Tttaken in Seconds

Dataset	SVM Model	Naïve Bayes Model	LSTM Model
100	1.063	0.320	0.092
200	2.686	1.811	1.074
300	3.968	3.792	2.904
400	4.232	3.132	2.324
500	5.325	4.897	4.124
600	6.984	6.788	6.563
700	7.570	7.184	6.632

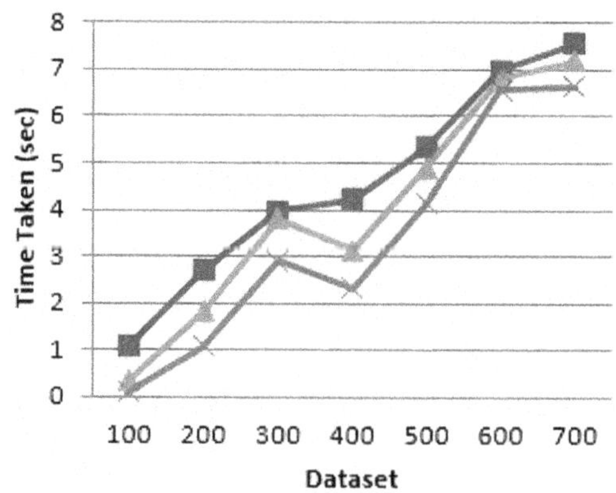

FIGURE 1.13 Comparison of time taken.

1.9 CONCLUSION

In the present research, simulation of conventional deep learning models has been done in order to identify and predict chronic disease. The dataset has been trained using a deep learning model in which classification is made using SVM, Naïve classifiers, and the LSTM model. In these simulations, it is concluded that LSTM is more accurate of better than SVM and Naïve classifiers, whereas the LSTM model is less time consuming and produces comparatively fewer errors rate than the SVM and Naïve models. It is concluded that the error rate of the LSTM model is 40% less than SVM and 20% less than the Naive Bayes model. Moreover, time consumption of the LSTM model is almost 50% of the SVM model and 25% of the Naive Bayes model.

1.10 FUTURE SCOPE

In the future, we want to apply our suggested approach to the prediction of more chronic illnesses. Prediction of various chronic illnesses may benefit in particular from ensemble feature selection methods. Little work has been published in the field of developing a system that can identify different illnesses in patients, making it an open research challenge. Limitations of the research are lack of flexibility and scalability. Future research might be more scalable and flexible. More and more people throughout the world are coping with many chronic illnesses at once. To aid in the diagnosis of a patient with many chronic conditions, we want to construct a prediction model that makes use of categorization and feature selection techniques. The study's results may not generalize to a dataset with patients suffering from a wider variety of illnesses due to the fact that all categorization techniques were applied to a subset of disorders. While the results of our AI-augmented prediction model have been encouraging, many questions need to be answered. To begin, there are not yet any illness prediction systems that can be used on a worldwide scale. Second, the restricted data availability and various sets of variables employed in research studies mean that prediction performance does not scale from research to real-world clinical applications. Third, it is difficult to clinically validate prediction models and execute them in real time. Prediction results' improvement based on clinical validation is another unexplored field, as is the diagnosis of previously identified illnesses.

REFERENCES

[1] J. Liu, Z. Zhang, and N. Razavian, "Deep EHR: Chronic disease prediction using medical notes," 2018, [Online]. Available: http://arxiv.org/abs/1808.04928.

[2] D. Jain and V. Singh, "Feature selection and classification systems for chronic disease prediction: A review," *Egypt. Inform. J.*, vol. 19, no. 3, pp. 179–189, 2018, doi: 10.1016/j.eij.2018.03.002.

[3] P. Kotturu, V. V. S. Sasank, G. Supriya, C. S. Manoj, and M. V. Maheshwarredy, "Prediction of chronic kidney disease using machine learning techniques," *Int. J. Adv. Sci. Technol.*, vol. 28, no. 16, pp. 1436–1443, 2019, doi: 10.17148/IJARCCE.2018.71021.

[4] J. Stewart, P. Sprivulis, and G. Dwivedi, "Artificial intelligence and machine learning in emergency medicine," *EMA Emerg. Med. Australas.*, vol. 30, no. 6, pp. 870–874, 2018, doi: 10.1111/1742-6723.13145.

[5] D. R. Rizvi, I. Nissar, S. Masood, M. Ahmed, and F. Ahmad, "An LSTM based deep learning model for voice-based detection of Parkinson's disease," *Int. J. Adv. Sci. Technol.*, vol. 29, no. 5 (Special Issue), pp. 337–343, 2020.
[6] R. Katarya and P. Srinivas, "Predicting heart disease at early stages using machine learning: A survey," *Proc. Int. Conf. Electron. Sustain. Commun. Syst. ICESC 2020*, no. Icesc, pp. 302–305, 2020, doi: 10.1109/ICESC48915.2020.9155586.
[7] M. Y. Shaheen, "Adoption of machine learning for medical diagnosis," *Sci. Prepr.*, no. September, pp. 0–2, 2021, doi: 10.31219/osf.io/t69w5.
[8] D. K. Plati et al., "A machine learning approach for chronic heart failure diagnosis," *Diagnostics*, vol. 11, no. 10, pp. 1–15, 2021, doi: 10.3390/diagnostics11101863.
[9] R. W. Pettit, R. Fullem, C. Cheng, and C. I. Amos, "Artificial intelligence, machine learning, and deep learning for clinical outcome prediction," *Emerg. Top. Life Sci.*, vol. 5, no. 6, pp. 729–745, 2021, doi: 10.1042/ETLS20210246.
[10] A. I. Chowdhury and K. A. Mamun, "An exploration of machine learning and deep learning-based diabetes prediction techniques," no. September, pp. 277–285, 2023, doi: 10.1007/978-981-19-4676-9_23.
[11] C. Mondol et al., "Early prediction of chronic kidney disease: A comprehensive performance analysis of deep learning models," *Algorithms*, vol. 15, no. 9, 2022, doi: 10.3390/a15090308.
[12] S. Srivastava, R. K. Yadav, V. Narayan, and P. K. Mall, "An ensemble learning approach for chronic kidney disease classification," *J. Pharm. Negative Res.*, vol. 13, no. 10, pp. 2401–2409, 2022, doi: 10.47750/pnr.2022.13.S10.279.
[13] D. A. Debal and T. M. Sitote, "Chronic kidney disease prediction using machine learning techniques," *J. Big Data*, vol. 9, no. 1, 2022, doi: 10.1186/s40537-022-00657-5.
[14] G. Chen et al., "Prediction of chronic kidney disease using adaptive hybridized deep convolutional neural network on the internet of medical things platform," *IEEE Access*, vol. 8, pp. 100497–100508, 2020, doi: 10.1109/ACCESS.2020.2995310.
[15] K. Balabaeva and S. Kovalchuk, "Comparison of temporal and non-temporal features effect on machine learning models quality and interpretability for chronic heart failure patients," *Procedia Comput. Sci.*, vol. 156, pp. 87–96, 2019, doi: 10.1016/j.procs.2019.08.183.
[16] M. Fabietti et al., "Artifact detection in chronically recorded local field potentials using long-short term memory neural network," *14th IEEE Int. Conf. Appl. Inf. Commun. Technol. AICT 2020 Proc.*, 2020, doi: 10.1109/AICT50176.2020.9368638.
[17] G. Battineni, G. G. Sagaro, N. Chinatalapudi, and F. Amenta, "Applications of machine learning predictive models in the chronic disease diagnosis," *J. Pers. Med.*, vol. 10, no. 2, 2020, doi: 10.3390/jpm10020021.
[18] T. H. H. Aldhyani, A. S. Alshebami, and M. Y. Alzahrani, "Soft clustering for enhancing the diagnosis of chronic diseases over machine learning algorithms," *J. Healthc. Eng.*, vol. 2020, 2020, doi: 10.1155/2020/4984967.
[19] T. Zhu, K. Li, P. Herrero, and P. Georgiou, "Deep learning for diabetes: A systematic review," *IEEE J. Biomed. Heal. Inform.*, vol. 25, no. 7, pp. 2744–2757, 2021, doi: 10.1109/JBHI.2020.3040225.
[20] L. Men, N. Ilk, X. Tang, and Y. Liu, "Multi-disease prediction using LSTM recurrent neural networks," *Expert Syst. Appl.*, vol. 177, no. March, p. 114905, 2021, doi: 10.1016/j.eswa.2021.114905.
[21] X. Luo et al., "Applying interpretable deep learning models to identify chronic cough patients using EHR data," *Comput. Methods Programs Biomed.*, vol. 210, p. 106395, 2021, doi: 10.1016/j.cmpb.2021.106395.
[22] K. Divya Vani, "Heart disease prediction using long short-term memory (LSTM) deep learning methodology," *Int. J. Res. Publ. Rev.*, vol. 2, no. 8, pp. 1076–1081, 2021, [Online]. Available: www.ijrpr.com.

[23] N. Bhaskar, M. Suchetha, and N. Y. Philip, "Time series classification-based correlational neural network with bidirectional LSTM for automated detection of kidney disease," *IEEE Sens. J.*, vol. 21, no. 4, pp. 4811–4818, 2021, doi: 10.1109/JSEN.2020.3028738.
[24] R. S. Segall and V. Sankarasubbu, "Survey of recent applications of artificial intelligence for detection and analysis of COVID-19 and other infectious diseases," *Int. J. Artif. Intell. Mach. Learn.*, vol. 12, no. 2, pp. 1–30, 2022, doi: 10.4018/ijaiml.313574.
[25] A. A. Nancy, D. Ravindran, P. M. D. Raj Vincent, K. Srinivasan, and D. Gutierrez Reina, "IoT-cloud-based smart healthcare monitoring system for heart disease prediction via deep learning," *Electronics*, vol. 11, no. 15, 2022, doi: 10.3390/electronics11152292.
[26] Y. Kumar, A. Koul, R. Singla, and M. F. Ijaz, "Artificial intelligence in disease diagnosis: A systematic literature review, synthesizing framework and future research agenda," *J. Ambient Intell. Humaniz. Comput.*, 2022, doi: 10.1007/s12652-021-03612-z.
[27] H. Chen, M. Du, Y. Zhang, and C. Yang, "Research on disease prediction method based on R-lookahead-LSTM," *Comput. Intell. Neurosci.*, vol. 2022, 2022, doi: 10.1155/2022/8431912.
[28] M. Badawy, N. Ramadan, and H. A. Hefny, "Healthcare predictive analytics using machine learning and deep learning techniques: A survey," no. Ml, pp. 1–44, 2022, [Online], doi: 10.21203/rs.3.rs-1885746/v2.
[29] R. Alanazi, "Identification and prediction of chronic diseases using machine learning approach," *J. Healthc. Eng.*, vol. 2022, 2022, doi: 10.1155/2022/2826127.
[30] M. M. Ahsan, S. A. Luna, and Z. Siddique, "Machine-learning-based disease diagnosis: A comprehensive review," *Healthcare*, vol. 10, no. 3, 2022, doi: 10.3390/healthcare10030541.
[31] G. Kim, H. Lim, Y. Kim, O. Kwon, and J. H. Choi, "Intra-person multi-task learning method for chronic-disease prediction," *Sci. Rep.*, vol. 13, no. 1, pp. 1–10, 2023, doi: 10.1038/s41598-023-28383-9.
[32] P. Anandajayam, S. Aravindkumar, P. Arun, and A. Ajith, "Prediction of chronic disease by machine learning," *2019 IEEE Int. Conf. Syst. Comput. Autom. Netw. ICSCAN 2019*, pp. 1–6, 2019, doi: 10.1109/ICSCAN.2019.8878724.
[33] Z. Li, J. Huang, and Z. Hu, "Screening and diagnosis of chronic pharyngitis based on deep learning," *Int. J. Environ. Res. Public Health*, vol. 16, no. 10, 2019, doi: 10.3390/ijerph16101688.
[34] C. Kim, Y. Son, and S. Youm, "Chronic disease prediction using character-recurrent neural network in the presence of missing information," *Appl. Sci.*, vol. 9, no. 10, 2019, doi: 10.3390/app9102170.
[35] A. Varol and Institute of Electrical and Electronics Engineers, "1st international informatics and software engineering conference (IISEC-2019) : 'Innovative technologies for digital transformation' : Proceedings book : 6–7 November 2019, Ankara/Turkey," no. 2, pp. 1–4, 2019. IEEE.
[36] S. T. Himi, N. T. Monalisa, M. D. Whaiduzzaman, A. Barros, and M. S. Uddin, "MedAi: A smartwatch-based application framework for the prediction of common diseases using machine learning," *IEEE Access*, vol. 11, no. October 2022, pp. 12342–12359, 2023, doi: 10.1109/ACCESS.2023.3236002.
[37] V. D. Soni, "Chronic disease detection model using machine learning techniques," *Int. J. Sci. Technol. Res.*, vol. 9, no. 09, pp. 262–266, 2020.

2 A Novel Approach to Multimedia Malware Detection Using Bi-LSTM and Attention Mechanisms

Tanisha Bansal, Kiran Malik, Shambhu Sharan, Poonam Bansal

2.1 INTRODUCTION

Malware, an abbreviation for malicious programs, refers to any program or code, including viruses, Trojans, worms, spyware, and so on, that tends to cause harm to computing devices like computers, servers, or networking equipment [1]. Internet technology has permeated many facets of our life as digital technology has advanced. The internet has provided individuals with unparalleled convenience and numerous prospects for the advancement of emerging technology; however, it has created more critical network security issues too. Numerous focused assault groups have utilised malware to harm as well as impair commercial activities in recent years [2–4]. Every year, the number of malware assaults on personal computers, cell phones, and other smart devices increases significantly. Among the most significant studies in the world of cyberspace security has been the investigation of proper and effective multimedia malware technologies. Signature-based and heuristic detection are the most often utilised classical detection approaches, and the first is the most extensively employed approach [5, 6]. This approach generates a signature database by analysing prior well-identified malware and extracts its signature characteristics, then compares the characteristics of suspected programs to be recognised in relation to a database so as to identify the malware. This strategy functions effectively for existing malware but not for newly discovered malware.

The other sort of heuristic detection involves detecting samples based on criteria developed by expert researchers, which is less efficient and significantly reliant on individual expertise [7]. With the rapid increase in the quantity and varieties of malware, as well as a massive number of undiscovered forms of malware, conventional approaches are clearly unable to deal with the issue. To effectively deal with the prevailing concern of malware expansion, competent investigators have focused on malware identification techniques using machine learning (ML) [8]. However, customary

DOI: 10.1201/9781003477280-2

ML detection techniques even now depend on conventional design and involvement in the feature extraction step, which further necessitate fulfil a set of assumptions, and there are predetermined restrictions in such a feature extraction, which ultimately affect the accuracy of detection. Deep learning (DL), on the other hand, can acquire additional knowledge and relationships using the same inputs or characteristics; hence, malware detection using DL is increasingly becoming a new research focus [9]. While a few improvements have been made, there are certain concerns that require additional investigation at this time. For instance, the majority of the present work makes use of learning models based mostly on convolutional neural networks (CNN), wherein those models may be subjected to a variety of experiments. For instance, recurrent neural networks (RNN) are adept at managing information by means of sequencing like byte/opcode sequences, whereas graph neural networks (GNN) are mainly proficient in managing multiple kinds of call graph functionalities like function call graphs, control flow, data flow, and so on. However, owing to the model's dependent parameters, there are restrictions on the choice as well as the evaluation of sample characteristics in this sort of study; thus, malware behaviour is not fully learnt.

As the types of malware grow in number, the static properties will change, rendering such solutions ineffective for coping against unidentified malware, and there is potential for ongoing advancement in this domain. In comparison to static assessment, dynamic assessment can assist in the identification of unidentified malicious programmes by observing and evaluating the actual behaviour of the code execution. Because comparable attack intent typically correlates to certain common combinations of similar actions, dynamic assessment might accurately represent the code's true intent and behavioural insights. This research provides a malicious programme-detecting model having an attention mechanism alongside Bi-LSTM to address the aforementioned issues. We choose the API calls acquired through dynamic analysis to appropriately depict the behavioural features of the samples. To identify malware, an attention function is also added into the detection process. To aid MDM training, we acquired around 16k of Windows malware from different websites like virusshare, as well as approximately 11k of regular apps from Windows-based system and websites, to create a malware identification database. MDM outperforms prior approaches in all indicators in experimental testing, demonstrating that our technology is beneficial for enhancing the detection of the malware.

2.2 LITERATURE REVIEW

Conventional malware detection tools often employ static or dynamic analysis methodologies. The software is not required to execute during static analysis. It initially acquires the needed features by reversed code examination, subsequently uses human knowledge to construct the malware profile library needed for identification, and lastly compares to actually finish the identification. Since static-based evaluation techniques rely heavily on expert levels of knowledge, it's quite hard to observe robustly. There is a phenomenon known as underreporting, and the greatest dynamic analysis systems can only recognise malicious code with an established fingerprint catalogue but cannot find malicious activity not available in the fingerprint catalogue. Dynamic approach–based evaluation determines the underlying properties of

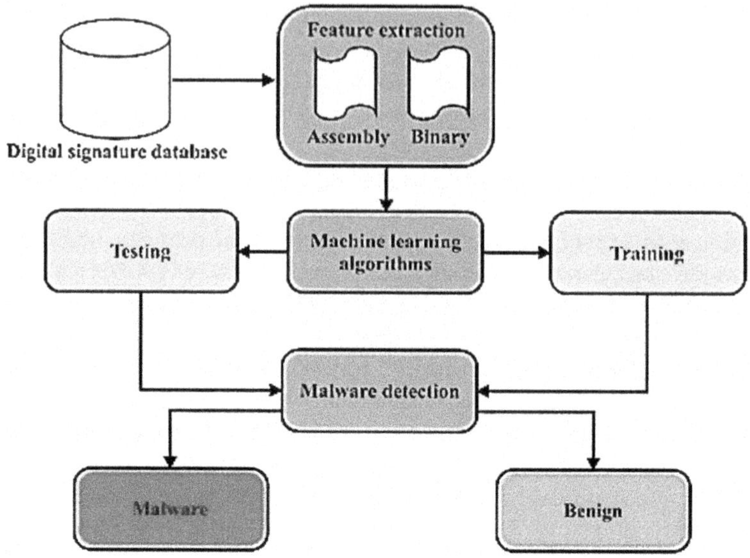

FIGURE 2.1 Overall cycle of intelligent malware detection [10].

a programme by running it in a simulated space such as docker and analysing its contents during runtime. Dynamic analysis, as opposed to static analysis, may examine the functionality of the program more explicitly. Conventional detection approaches are nevertheless unable to deal with both static and dynamic analysis in light of the quick expansion of huge amounts of undiscovered malware. With the widespread use of AI technology, the majority of new malware identification technique development is focused on machine learning (ML) and artificial intelligence (AI), and we refer to this type of work as the intelligent malware detection approach. The usual flow of this sort of job is seen in Figure 2.1.

The method consists of the following steps: first, attempting to determine a dataset; second, trying to analyse codes using some analysis method, either static or dynamic, and trying to harvest essential functionality; third, handling the attributes, particularly regarding vectorising the attributes to end up making the features meet the requirement specification of an eventual framework; and ultimately, attempting to build a prototype of comparable detection to determine if a specimen to be identified is malware or not. Existing research on intelligent malware detection has some successes in malicious code discovery and group categorisation. The majority of early research focused on static aspects like opcode/byte sequences, strings, headers of programmes, and so on. A research section also exists that translates programme data into picture format for detection and analysis. Assaleh et al. used the K-mean nearest neighbour (KNN) method to identify 65 portable executable files relying on the N-gram property of the byte sequence, with a 98% accuracy rate [11]. The resulting strings in the PE files were retrieved employing four distinct detection algorithms, and the results demonstrate that the SVM has the best accuracy. Shafiq et al.

collected 189 characteristics from the PE header, which were then processed using feature engineering and utilised for further detection [12]. The DLL and header properties of the program to be examined were used as an input mean for the classification model by Li et al. [13]. Many more studies based on classic ML techniques have been presented, including decision trees (DT), random forests (RF), support vector machines (SVM), Naive Bayes (NB), and others.

Deep learning (DL) techniques, as opposed to typical ML algorithms, can learn characteristics autonomously, which may increase detecting performance and reduce false positive rates, thus becoming one among the major fields academically for identifying malicious programs. In the present work, the most often utilised DL network models are convolutional neural networks (CNN) and recurrent neural networks (RNN). Numerous works have been carried out by way of CNN models in identification work on malware that convertsd programs into picture for processing and evaluation: i.e. byte values of executables are aligned to image pixels by way of different sets of rules, and the resulting pictures are identified by image recognition techniques to accomplish the tasks of malicious program detection [14–16]. Additionally, few RNN-based models build on the effective implementation of RNNs in natural language processing (NLP). This sort of work takes into account the resemblances among computer code and natural language and often transforms recovered sequential characteristics to word vectors to input RNN and its many types of networks to finish the identification and categorisation task [17, 18, 20]. In the beginning, the sequence characteristics were primarily static, including opcode sequences and byte sequences, but as the study progressed, various dynamic aspects began to be included, taking into account the time-based connection of the actions inside the specimens [19]. One of the publications then discusses static code analysis shortcomings in evaluating obfuscated sample data and suggests that dynamic analysis approaches are better suited for harmful software analysis [21].

2.3 METHODOLOGY

The characteristics generated by dynamical assessment and deep learning (DL) detection techniques are the subject of this study. This section describes our methodology in detail. The goal of our effort is to recognise malicious software that is primarily a binary classification problem. We concentrate on the execution time characteristics of Windows programmes to make a more accurate evaluation. Because API calls all through the program execution describe the application's behaviour and true intent, we capture the API call sequences of the running application as attributes and use hidden knowledge to determine if the instance undergoing analysis is malicious. Though distinct malwares' individual assault intentions and API call mechanisms might not be identical, the core APIs in the API call sequencing of diverse attack intentions are conceptually comparable. To correctly collect the semantic and temporal aspects of the API for the purpose of increasing the accurateness level of detection, we used the bi-directional LSTM and incorporated the attention function, thereby creating a malware detection model (MDM). The MDM system is made up of three components: input preprocessing, neural network (NN)–based training and optimisation, and unknown sample identification.

2.3.1 Input Preprocessing

Windows pre-installation environment (WinPE) files cannot be utilised straightforwardly as NN inputs. The data must be preprocessed correctly in order to train the neural network. The separation of API call sequences and word embedding encoding are performed consecutively in the sample preprocessing stage. To depict program activity, we utilise Cuckoo Sandbox to collect example API call patterns. Initially, we separate the API call records of the malicious programmes for every case using different sandbox analysis tools like Cukoo and BSA, and subsequently, we do API call pattern analysis on the gathered API call records. We shorten the files calling five thousand or more APIs in a single thread and don't de-duplicate any sequences of API to gather as many comprehensive behavioural and temporal details as feasible while avoiding the challenge of substantial training cost caused by extended sequences. Because one or more call sequences of the observations can't be utilised straightforwardly as direct inputs to NN, we perform vectorisation on them by means of a strategy akin to the word vector model in NLP.

2.3.2 Neural Network–Based Training and Optimisation

The classic RNN trains attributes across time using the back-propagation technique and the gradient descent, limiting the network's capacity to gain knowledge of extended statistical data. In our study, the LSTM module is employed to substitute the RNN's hidden component. We utilise the Bi-LSTM structure to understand the features since it can concurrently use the contextual data of the present state for learning purposes and is generally more proficient in retrieving attributes than the unidirectional LSTM. In an attempt to acquire long-term interrelated characteristics more successfully and to dwell on crucial info, we included an attention embedding in the system-building phase. As a result, we developed a NN system based on Bi-LSTM having the attention method incorporated, utilised the outcome of the preceding stage's embedding layer as the input of the detecting model for training, and employed Adam Optimizer to optimise the NN model's hyper-parameters to achieve a superior version of modeling.

2.3.2.1 Bi-LSTM Layer

This is an upgraded version of long short-term memory that includes both forward and backward memory. An LSTM cell is present in every temporal phase to effectively remember, forget, and output relevant data. Assume you have a malicious API sequence information S comprising T items, and the same is represented as $S = [x1, x2, ..., xT]$. The embedding layer converts each xi into a real vector ei, wherein ei is the ith item (D-dimensional embedding) in the series, and E is a matrix of T x D dimension constructed by the embedding depictions of all items in the series. When E is fed into the Bi-LSTM system, the forward (fw) LSTM's hidden state output $\overrightarrow{h_t}$ is represented in (1) while the backward LSTM's hidden state output is $\overleftarrow{h_t}$ (outlined in (2)):

$$\overrightarrow{h_t} = \overrightarrow{LSTM}\left(e_i, \overrightarrow{h_{t-1}}\right) \tag{2.1}$$

$$\overleftarrow{h_t} = \overleftarrow{LSTM}\left(e_i, \overleftarrow{h_{t+1}}\right) \tag{2.2}$$

The outcome of the hidden state of Bi-LSTM at interval t is combined by $\overrightarrow{h_t}$ and $\overleftarrow{h_t}$, that might be represented as $h_t = [\overrightarrow{h_t}, \overleftarrow{h_t}]$. So the hidden set of states represented by $H = [h1,\ h2,\ \ldots,\ hT]$ is achieved after the sequential learning reaches to an end. Presuming the possible unit of nodes that are hidden in the Bi-LSTM is U, the dimensionality of the hidden state set is $T \times 2U$. The system's primary objective is to transform sequences of varying lengths to fixed-length encoding depictions using the sequential weighted totaling of the T hidden states of the LSTM. The weight assignment estimation necessitates the adoption of the attention method.

2.3.2.2 Concept of Attention

The attention mechanism uses the hidden set of states – i.e. H of the Bi-LSTM – as input, performs nonlinear transformation using the activation function Tanh, realises dimensional reduction, and afterwards normalises with SoftMax to produce the final attention vector. The SoftMax mechanism ensures that every member of the resultant attention vector reflects a likelihood, so the total of over all components equals one. Equations (3) to (6) show the weight matrices created by the attention layer. Here, T is size of sentence, H ∈ RD x T, and D is the word vector size. The dimension is D, and wT is the transposed parameter vector learnt during training, and r is the vector form of the API sequence. So if h* is transferred to the dense layer, the attention layer output data is dropped out to minimise the fitting problem before being sent to the completely linked layer, where a SoftMax classification system is used to identify the observed sample and report the detected findings.

$$\mathrm{M} = \tanh(\mathrm{H}) \tag{2.3}$$

$$\alpha = \mathrm{softmax}\left(\mathrm{w}^{\mathrm{T}}\mathrm{M}\right) \tag{2.4}$$

$$\mathrm{r} = \mathrm{H}\alpha^{\mathrm{T}} \tag{2.5}$$

$$\mathrm{h}^{*} = \tanh(\mathrm{r}) \tag{2.6}$$

2.3.2.3 Module for Detecting Unknown Samples

Training and optimisation are conducted following the collection of the API sequence from the targeted specimen to be identified and conducting word embedding processing, and the detecting system is built to determine if the target program is malicious or a harmless specimen.

2.4 RESULTS AND DISCUSSION

In this part of the chapter, we present the empirical arrangement and dataset, then subsequently undertake MDM operational analysis and evaluation.

2.4.1 Basic Configuration

Table 2.1 shows the physical and logical environmental setup for the chapter's methodical deployment based on MDM.

TABLE 2.1
Setup of the MDM System's Software/Hardware Environment

Parameter	Specification
CPU	Dual AMD Rome 7742, 128 cores total, 2.25 GHz (base), 3.4 GHz (max boost)
GPU	8x NVIDIA A100 80 GB/40 GB Tensor Core GPUs
OS	Linux – Ubuntu
Software Framework	Docker

Because the API calls must be retrieved by dynamically executing the programme, binary forms of the files are necessary. As a result, the set of data utilised in our research was created by us. Websites like malicia-project.com and virusshare.com were used to acquire all malicious sample programs. Winwebsec, Locker, Zbot, Zeroaccess, and Mediyes are among the malware that were discovered, totaling around 16k files. VirusTotal validates all files to ensure they are the correct kind. Malicious executables are gathered from the installation directories of several sorts of genuine application programs. The harmless application collection consists of two components, the first obtained from Windows computers while the second was retrieved from the link www.cnet.com/windows/website, for a total of approximately 11k sample programs. We shortened files that access over 5,000 APIs in a single thread in one file due to the time complexity and effectiveness of model verification.

2.4.2 Assessment and Metrics for Model Evaluation

We consistently classify malicious specimens as viruses and harmless ones as benign. To assess the correctness of the system, we employ a ten-fold cross-validation procedure. In other words, the set of data is broadly assigned into ten parts, with nine of them serving as training examples, and one serving as testing data for substantiation. We conducted this research ten times, and the findings were normalised. To assess the success of our system, we chose accuracy, precision, recall, and F1 measure from a list of regularly used assessment parameters in the malware recognition domain. Table 2.2 depicts the categorisation problem's confusion matrix, wherein TP denotes true positive (i.e. the number of rightly estimated positive samples); FP represents false positive (i.e., the number of negatively predicted items as positive items); FN denotes the number of positively predicted specimens identified as negative; and TN denotes the number of rightly determined negative items.

The number of successfully categorised data items divided by the entire number of data items is referred to as accuracy, as shown in equation (7).

$$Accuracy = \frac{TP + TP}{TP + FP + TN + FN} \tag{2.7}$$

Equation (8) defines the precision. An effective classifier may generally have a precision of one (high). Precision reaches one when the numerator and the denominator

TABLE 2.2
Binary Classifier's Confusion Matrix

	Actual Positive	Actual Negative
Predicted Positive	True Positive (TP)	False Positive (FP)
Predicted Negative	False Negative (FN)	True Negative (TN)

are identical (i.e. TP = TP + FP); this even implies that FP is zero. Whenever FP rises, the value of the denominator gets bigger than the value of the numerator, and the precision value drops.

$$Precision = \frac{TP}{TP + FP} \tag{2.8}$$

Recall, also referred as true positive or sensitivity rate, is defined as shown in equation (9). For an effective predictor, recall may typically be one (high). Whenever the numerator and denominator become equivalent (i.e. TP = TP + FN), recall become one; it also implies that FN is zero. When FN grows, the denominator value gets bigger than the numerator, and the value of recall gets lower.

$$Recall = \frac{TP}{TP + FN} \tag{2.9}$$

The F1 score is a statistic that considers both accuracy and recall and thus is expressed as shown in equation (10). Only when accuracy and recall are both one does the F1 score become one. Only when both accuracy and recall are strong can the F1 score rise. The F1 score is a preferable metric than accuracy since it is the harmonic mean of recall and precision.

$$F1 = \frac{2\,x\,precision + recall}{precision + recall} \tag{2.10}$$

In the preceding equation, TP indicates the number of malicious programs accurately detected in this study; FN reflects the amount of malware deemed harmless by the classifier. FP represents false positives, or the number of harmless specimens wrongly classified as malicious; TN represents the number of harmless specimens successfully recognised. It can be observed that accuracy and precision are taken into account in conjunction with TP and FP. The proportion of malicious programs which might be successfully recognised as a percentage of overall malware is represented by recalls. We did not utilise the false positive rate (FPR) since it counts the fraction of safe items that are misdiagnosed as malicious programs. However, in actuality, it is much more harmful to recognise malicious files as innocuous programs; thus, the recall rate ought to be especially considered in the issue of vulnerability detection. Accuracy and recall have an inverse relationship; as precision raises, recollection declines, and vice versa. If the proportion of malware-containing items in the test set is zero, the recall rate is one; however, the precision rate is quite low. As a result, F1 and its range of values are used to weigh the link between accuracy and recall. F1 is typically between [0,1], with greater F1 indicating better model performance.

2.4.3 Findings of the Analysis

This section describes our assessment results to illustrate the usefulness of MDM. We use a classification system to detect various max durations of API request sequences. Since the series of API calls is small (for instance, fewer than 50), it is unable to capture all conceivable harmful or typical activities; if the series is somewhat lengthy, more interruption and costs will be encountered. Furthermore, the greater the number of neurons contained in every hidden layer, the greater the model's time complexity. We ended up choosing the upper limit sequence length of 400, the count of hidden units, as well as the encoding dimensions of 256 as characteristics for additional learning and enhancement of the recognition model upon trying out varied dimensions and numbers of hidden units, especially in combination with the test findings, and exhaustively evaluating the time cost and the model's integration pace. The dropout approach is employed to limit fitting problem, whereas cross-entropy and Adam Optimizer loss functions have been adopted in this design. The cross-entropy can be represented as shown in equation (11). Between these, yi is the matching element in the SoftMax-normalised output vector, and yi' is the label's i-th value. It can be observed that whenever the categorisation is relatively precise, the associated element of yi would be nearer to one, and the associated element of Hyi'(y) would be nearer to zero. After computing the cross-entropy of every individual item, aggregate the cross-entropy of individual items to determine the loss value.

$$H_{y'}(y) = y_{i'}\log(yi) \tag{2.11}$$

The development trajectories of the four evaluation metrics throughout the training procedure are shown in Figure 2.2 when the value of epoch is selected as 20.

We contrast our solution to other API-based malicious program detection methods. Reference [22] employed the GIST global feature and Gabor wavelet transform–based technique to recognise and classify malicious files with around 96% accuracy using a feed-forward ANN. The frequency of API requests is chosen as a characteristic to identify malware. Their approach is to correlate the APIs utilised by every instance to a repository and then compute the frequency of API calls made by malicious programmesmes. The combination classifier detection using KNN and SVM yields 93% accuracy. There are several works that are comparable; we will not include them all here.

Incorporating these approaches, we utilise K-nearest neighboursn (KNN), artificial neural network (ANN), random forest (RF), support vector machine (SVM), and models to assess the efficiency of our method, and the analysis findings are displayed in Table 2.3.

According to the testing outcomes, the MDM's accuracy is 97.6%, approximately 2.3% percent better than alternative systems, whereas precision and F1 measure metrics are 98.8% and 98.7%, respectively, which are superior to traditional methods

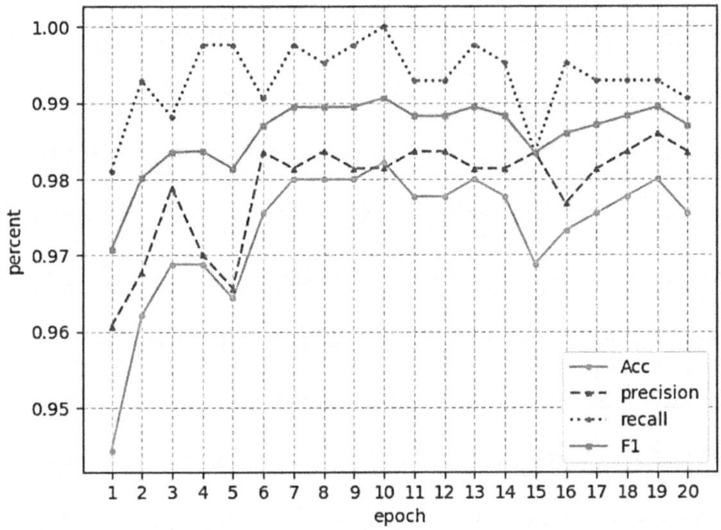

FIGURE 2.2 Evaluation metrics measurement variation.

TABLE 2.3
Evaluation of Various Approaches

Model	Accuracy (%)	Precision (%)	Recall (%)	F1 Measure (%)
KNN	94.3	93.7	94.5	94.1
ANN	95.3	95.8	95.1	95.5
RF	91.7	91.8	91.7	91.7
SVM	94.7	94.3	96.6	93.9
MDM	97.6	98.8	98.6	98.7

based on these ML algorithms. Furthermore, MDM has a recall rate of 98.6%, which is approximately 3.5%, 4.1%, 2%, and 6.9% better than MLP, KNN, SVM, and RF, respectively. The advancement is significant. In contrast to previous research focused on the FPR statistic, we believe it is more important to consider if the recall is enhanced. As FPR measures the fraction of harmless specimens incorrectly labelled as malicious programs, the real harm caused by malicious code being mislabeled as harmless data would be greater in this case. As a result, improving recall has a significant practical benefit for malware detection systems.

2.5 CONCLUSION

We suggested a malware detection model (MDM) based on Bi-LSTM in conjunction with attention mechanism in this chapter. The weight of the sequence characteristics of the malware API calls is calculated using the attention method, and the extraction of major characteristics is prioritised in the detection phase of the model. To put our strategy to the test, we acquired around 16k of malware samples from websites like virusshare and paired them with harmless software samples to create a dataset. When compared to previous studies, the test findings reveal that the proposed model has enhanced detection accuracy, precision, and recall, demonstrating the usefulness of our model. In our further research, we will strive to incorporate new models that function well enough in NLP, including such BGRU and Transformer, for sequence characteristics such as API call sequence and system call sequence, in order to increase the malware detection impact. Furthermore, we will continue to gather malware samples from more categories and harvest increasingly diverse sorts of characteristics in order to investigate malware identification and classification more thoroughly and efficiently.

REFERENCES

[1] R. Tahir, "A study on malware and malware detection techniques," *Int. J. Educ. Manag. Eng.*, 2018, doi: 10.5815/ijeme.2018.02.03.

[2] A. Shalaginov, K. Franke, and X. Huang, "Malware beaconing detection by mining large-scale DNS logs for targeted attack identification," *Int. J. Comput. Electr. Autom. Control Inf. Eng.*, vol. 10, 2016.

[3] L. E. Branch et al., "Trends in malware attacks against United States healthcare organizations, 2016–2017," *Glob. Biosecur.*, 2019, doi: 10.31646/gbio.7.

[4] Ş. Bahtiyar, "Anatomy of targeted attacks with smart malware," *Secur. Commun. Netw.*, 2016, doi: 10.1002/sec.1767.

[5] R. Arora, A. Singh, H. Pareek, and U. R. Edara, "A heuristics-based static analysis approach for detecting packed PE binaries," *Int. J. Secur. Appl.*, 2013, doi: 10.14257/ijsia.2013.7.5.24.

[6] S. K. Sahay, A. Sharma, and H. Rathore, "Evolution of malware and its detection techniques," in M. Tuba, S. Akashe, and A. Joshi (eds) *Information and Communication Technology for Sustainable Development. Advances in Intelligent Systems and Computing*, vol 933. Springer, Singapore, doi: 10.1007/978-981-13-7166-0_14

[7] S. Treadwell and Z. Mian, "A heuristic approach for detection of obfuscated malware," 2009, doi: 10.1109/ISI.2009.5137328.

[8] S. A. Habtor and A. H. H. Dahah, "Machine-learning classifiers for malware detection using data features," *J. ICT Res. Appl.*, 2021, doi: 10.5614/ITBJ.ICT.RES.APPL.2021.15.3.5.

[9] S. Ni, Q. Qian, and R. Zhang, "Malware identification using visualization images and deep learning," *Comput. Secur.*, 2018, doi: 10.1016/j.cose.2018.04.005.

[10] A. Souri and R. Hosseini, "A state-of-the-art survey of malware detection approaches using data mining techniques," *Human Centric Comput. Inf. Sci.*, 2018, doi: 10.1186/s13673-018-0125-x.

[11] T. Abou-Assaleh, N. Cercone, V. Kešelj, and R. Sweidan, "N-gram-based detection of new malicious code," 2004, doi: 10.1109/cmpsac.2004.1342667.

[12] M. Z. Shafiq, S. M. Tabish, F. Mirza, and M. Farooq, "PE-miner: Mining structural information to detect malicious executables in real time," 2009, doi: 10.1007/978-3-642-04342-0_7.

[13] B. Li, K. Roundy, C. Gates, and Y. Vorobeychik, "Large-scale identification of malicious singleton files," 2017, doi: 10.1145/3029806.3029815.

[14] K. He and D. S. Kim, "Malware detection with malware images using deep learning techniques," 2019, doi: 10.1109/TrustCom/BigDataSE.2019.00022.
[15] J. Yan, Y. Qi, and Q. Rao, "Detecting malware with an ensemble method based on deep neural network," *Secur. Commun. Netw.*, 2018, doi: 10.1155/2018/7247095.
[16] M. Abdelsalam, R. Krishnan, Y. Huang, and R. Sandhu, "Malware detection in cloud infrastructures using convolutional neural networks," 2018, doi: 10.1109/CLOUD.2018.00028.
[17] S. Tobiyama, Y. Yamaguchi, H. Shimada, T. Ikuse, and T. Yagi, "Malware detection with deep neural network using process behavior," 2016, doi: 10.1109/COMPSAC.2016.151; I. Kwon and E. G. Im, "Extracting the representative API call patterns of malware families using recurrent neural network," 2017, doi: 10.1145/3129676.3129712.
[18] B. Athiwaratkun and J. W. Stokes, "Malware classification with LSTM and GRU language models and a character-level CNN," 2017, doi: 10.1109/ICASSP.2017.7952603.
[19] B. Kolosnjaji, A. Zarras, G. Webster, and C. Eckert, "Deep learning for classification of malware system call sequences," 2016, doi: 10.1007/978-3-319-50127-7_11.
[20] A. Moser, C. Kruegel, and E. Kirda, "Limits of static analysis for malware detection," 2007, doi: 10.1109/ACSAC.2007.21.
[21] A. Makandar and A. Patrot, "Malware analysis and classification using artificial neural network," 2016, doi: 10.1109/ITACT.2015.7492653.
[22] V. Garg and R. K. Yadav, "Malware detection based on API calls frequency," 2019, doi: 10.1109/ISCON47742.2019.9036219.

3 Exploring Machine Learning Applications for Enhancing Security and Privacy in Multimedia IoT

A Comprehensive Review

Ruchika, Suman

3.1 INTRODUCTION

Smart devices can communicate with one another via the IoT network, allowing information to be transferred across several platforms simply and efficiently. Making full use of the possibilities of the internet, the current modification of numerous wireless technologies puts IoT as the subsequent breakthrough equipment. Smart systems, such as smart water, smart transportation, smart offices, smart retail, smart agriculture, and smart healthcare, are now being deployed in smart cities [1]. Several diverse open networks supported by technologies like radio frequency identification (RFID) and wireless sensor networks (WSN) are predicted to connect over 50 billion items to the internet, laptops, sensors, smartphones, and game supports [2]. The IoT comprises three paradigms: sensors, knowledge, and the IoT. The IoT is the next technological revolution since it can access all the internet has to offer [3].

Daily, an increasing number of software programs and digital gadgets create huge volumes of organized or semi-structured data. Sensors, gadgets, social networking apps, healthcare apps, and temperature sensors are all examples. This immense data production has resulted in the term BD [4]. Due to inefficiencies, large volumes of data, colloquially known as BD, cannot be stored, processed, or analyzed using typical database systems.

Even though the term BD has been utilized in the past, companies and information technology (IT) experts are just now getting acquainted with it [5]. BD studies, corresponding to the McKinsey Global Institute, represent the next leading edge for invention, competitiveness, and output [6]. In this sense, BD refers to a better database system tool for obtaining, saving, managing, and evaluating large data. In the Digital Universe study [7] the current wave of BD technologies and architectures intends to extract value from massive quantities and formats of heterogeneous data

DOI: 10.1201/9781003477280-3

by permitting high-velocity data gathering, discovery, and investigation. Similarly, past research divides BD into three categories: the source of the information, how it's analyzed, and how the results are presented. As the IoT grows, the world's data volume is expected to grow exponentially to 44 zettabytes (multiple of unit bytes that measures digital storage) by 2020, as shown in Figure 3.1.

The number of connected "things" is expected to reach 32 billion by 2020, with the IoT system's data contribution accounting for 10% of the digital universe [9]. Even though each company's and research institution's projections of the effect of the internet of things varies, all indicate an increasing trend. According to some estimations, the business might be valued at up to US$3.9 – 11.1 trillion by 2025. According to one university's research, data analytics and application services will account for 77% of IoT business potential by 2020 [10].

3.1.1 Security Challenges in IoT

Today's digital infrastructures face a host of threats. A smart TV, a home Wi-Fi network, and an organization's network are all vulnerable to hacking. The IoT is no exception. To make things worse, due to the open architecture that supports the internet of things, devices on the IoT are inherently susceptible. Corresponding to mobility, wireless, embedded use; diversity; and size are all aspects of the IoT that contribute to security and privacy problems [11]. For these reasons, the security needs of the IoT are unique, posing various new difficulties in information security. The taxonomy given in Table 3.1 depicts the security risks in IoT.

3.1.2 IoT Security and Privacy

IoT device growth has a significant impact on sectors such as automotive, aviation, and smart homes, as well as medical wearables and agriculture. Due to the widespread use of IoT, both consumers and businesses are concerned about its security and privacy [12]; others have called for further criteria and controls to be imposed on its use [13]. There is the possibility of privacy breaches and safety and health issues because of a failure of IoT settings, such as automotive crashes and theft of monetary value through IoT pacemakers and pipeline explosions [14], as shown in Figure 3.2.

As a consequence of both IoT security and privacy issues, most efforts have concentrated on strengthening perimeter defenses, such as firewalls, on the IoT infrastructure [15]; intrusion detection techniques; access control policies [16]; and software patches [17]. Certain IoT devices and programming frameworks have also been investigated. In the past, source code analysis was also utilized to protect IoT applications and devices. The bulk of previous research has focused on mobile phone security solutions. The context of an app may be used to enforce permissions via runtime prompts or by seeking permission from users through an interface [18].

3.1.2.1 Security

The IoT is particularly vulnerable to security risks due to its distinctions from more traditional computer equipment [19]. Many IoT devices are designed for mass use;

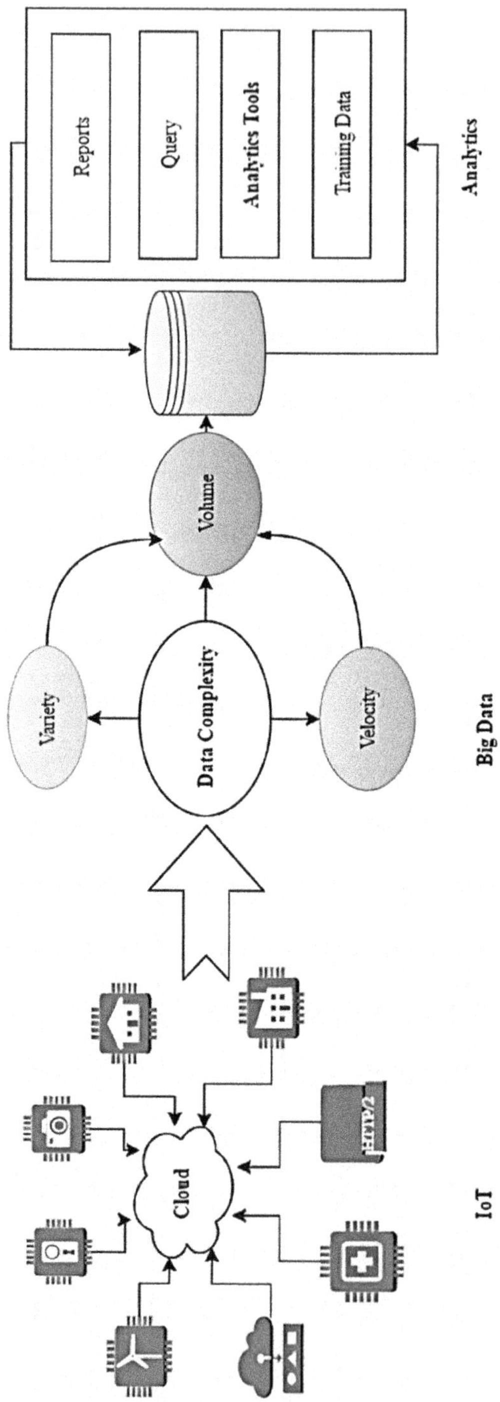

FIGURE 3.1 Correlation between IoT and BDA [8].

TABLE 3.1
Attacks in IoT

Top-Level Security Domain	Sub-Domains
Architecture	Perception Layer, Application Layer, and Network Layer
Threat Vector	Communication Attacks, Physical Attacks, and Application/Software Attacks
Trust	Privacy, Reliability, and Availability
Compliance	Policy control, Government Oversight, Non-Government Oversight

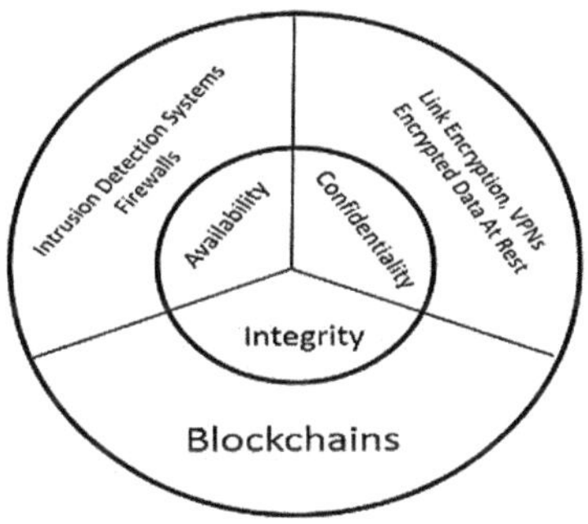

FIGURE 3.2 Security system for IOT.

smart systems and sensors are excellent instances of IoT. A typical IoT deployment consists of a collection of devices with comparable or nearly identical features. As they're all similar, a security problem that affects a huge number of individuals may be exacerbated [20]. Similarly, several organizations have produced risk assessment recommendations. As a result of this advancement, the number of networked IoT devices is on track to set a new high. Many of these devices are also capable of connecting to and interacting with other devices on their own in an erratic fashion. This needs a study of the widely accessible IoT security technologies, techniques, and tactics [21].

In terms of validation, IoT faces a variation of weaknesses, which remains among the most pressing problems in the provision of security in several applications. Denial of service (DoS) attacks and replay attacks are two instances of security concerns that the existing authentication technique does not sufficiently handle [22]. Due to the prevalence of hazardous apps in the IoT environment, data collection has become a significant concern and information security is one of the extremely susceptible areas in IoT validation: for instance, contactless credit cards [23]. Card numbers and names can be readily translated exclusive of IoT authentication, permitting hackers to buy products utilizing a bank account number and their identities. Third parties taking over a communication channel and using it to fake the identities of physical nodes involved in network exchanges as part of man-in-the-center attacks on the IoT is rare. The attacker does not need to know the details of a potential victim for the bank server to identify the transaction as legitimate [24].

3.1.2.2 Privacy

The value of the IoT is influenced by people's privacy choices. A lack of trust in IoT's capacity to preserve privacy and avoid potential harm might cause an impasse in its mainstream adoption [25]. It is vital to recognize that the IoT, connected devices, and the services that accompany them depend on users' confidence and faith in their privacy rights. With much effort, the IoT is redefining privacy problems, such as the surge in monitoring and tracking. The omnipresent intelligence-integrated artifacts of the internet of things, where sampling and information sharing can be done nearly everywhere, are posing privacy concerns. The fact that access to the internet is so ubiquitous that obtaining personal information from any area on the planet will be significantly easier if a unique approach is put in place is a crucial component in appreciating this problem [26].

3.1.2.3 Interoperability

It is well established that creating a disconnected, private IoT ecosystem reduces value for customers. Even though complete interoperability among goods and services is not always achievable, consumers might dislike buying goods and services that lack flexibility and are subject to dealer lock-in. If IoT devices are poorly designed, the networks to which they are linked can suffer [27]. Another critical feature is cryptography, which has long been applied to protect against security flaws in a variety of applications [28]. It is hard to provide a comprehensive defense against current attacks with only one type of security software. Therefore, many layers of security are required to protect against the threats that compromise IoT authentication [29]. By establishing more complicated security methods and implementing these elements in commodities, it is feasible to prevent hackers. Users may avoid this by buying things that already have enough safeguards in place to defend against possible security issues. Cybersecurity guidelines are among the procedures advocated to ensure the safety of the IoT [30].

3.1.3 Big Data and Its Analytics

Electronic devices (such as smart devices) and the internet era have had a tremendous influence on data computation [31]. IoT devices are used in remote sensing, software logging, wireless sensor networks, and mobile devices such as laptops and smartphones [32]. Consequently, zettabytes of BD are being generated, and the size is growing exponentially every year. BD can be described as an immense number of datasets, later, which incorporates other highlights [33]. BD's five Vs features are summarized and shown in Figure 3.3:

Volume: It implies a massive volume of data. Biological diversity can be developed in several ways. Therefore, a massive quantity of BD will be reviewed during analytics [34].

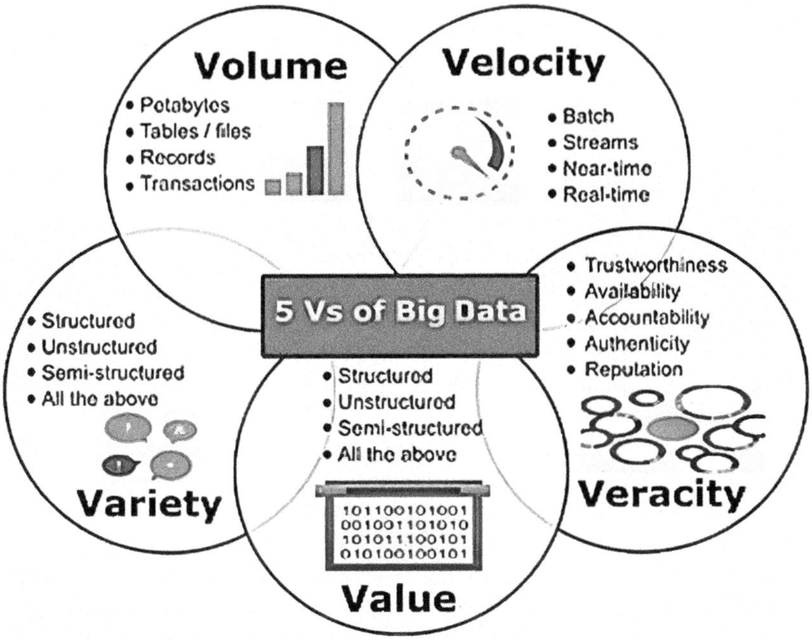

FIGURE 3.3 Five Vs paradigm of big data.

Variety: Structured, semi-structured, and unstructured BD formats are all included under this umbrella term. It is acceptable that the BD is in the form of a written document, a video or a picture, sensors, logs, and so forth. As BD gets more diversified, additional storage, mining, and analytics concerns emerge [35].

Velocity: BD develops because of an increase in the data production pace. It also covers the time spent analyzing data and making choices based on it [36].

Variability: It indicates the BD's instability, which implies that the data is always altering.

Veracity: It is, as the name implies, a measure of the reliability or quality of the data. Using business-driven truthfulness, unusual data is translated into insights into the dependability of BD.

3.2 LITERATURE OF REVIEW

This section contains a comprehensive review of literature on the statistical and data science challenges and opportunities: a study based on IoT security:

- Rehman et al., (2021) [37] stated that clinicians' decisions are becoming more proof based, which implies that no other sector has the same potential for analytics as healthcare. Due to the general amount and accessibility of healthcare data, analytics have transformed the healthcare business and opened up new possibilities. Early diagnosis, prediction, and prevention are just a few of the advantages that this technology may provide. As a result of BD's huge influence on healthcare, treatment costs have been reduced, and sickness diagnosis has improved. Healthcare professionals, patients, and researchers have all benefited from its usage in recent years since it has assisted them in better diagnosing and treating patients, as well as providing community care. Everyone in the healthcare industry can benefit from BDA. If BD leads to improved health outcomes, patients and caregivers must be actively included in the assessment and policy-making processes. To develop the foundation for BD analytics in healthcare, government agencies, healthcare professionals, hardware manufacturers, and pharmaceutical organizations, as well as people, data scientists, academia, and suppliers, must cooperate.

- Kumari et al., (2020) [38] stated that to increase the condition of life for its civilians while preserving a healthy environment, a smart city must have an intelligent infrastructure. An intelligent grid that uses information and communications technology (ICT) to boost the efficiency and security of energy generation, distribution, and consumption is becoming more important. Therefore, the smart grid (SG) helps the complete development of every smart city. Demand response and load assessment are only two of the numerous possible applications for the BD created by SG. Secure analytics from BD have proven to be game changers in a multitude of smart city applications, including SG and transportation. Many exploration disputes, such as safe data collecting and preprocessing protected load data managing and storing and capacity projection, are emphasized in a one-of-a-kind process model derived from a comprehensive taxonomy. Finally, a real-world example is offered to demonstrate the model's usefulness.
- Bhattarai et al., (2019) [39] explained that large volumes of data might be utilized to unlock previously unimagined possibilities for enhancing the electrical system in ways that benefit technology, society, and the economy. The advancement of power grid technology, as well as measuring and communication technologies, find an exceptional amount of heterogeneous BD. The most difficult challenges to solve when attempting to transform a massive diversified dataset into meaningful information are data security, computational complexity, and the operational combination of BD into power system planning and operational frameworks. In this scenario, BD analytics and visualization could increase situational awareness and predictive decision-making. The author analyzes data analytics and its purposes in power grids, offering an impression of the present state of the art in BD analytics. According to an issue of the *Journal of Business Analytics*, utility firms could establish new business patterns and profit flow by utilizing BD analytics in several ways. Consequently, the utility companies will be able to make the best financing and functioning outcomes at the right time and in the right location.

- Hassan et al., (2019) [40] explained that the discipline of healthcare informatics is undertaking a tidal shift with the low cost and broad availability of wearable sensors. Smart hospitals can now build remote patient monitoring (RPM) simulations that monitor patients from home thanks to IoT sensors. Using the AALs for long-term patient monitoring provides a lot of data. In the processing of huge medical datasets, Spark and its ML libraries are ten times quicker than MapReduce. Cloud technologies that can deal with huge volumes of data are opening the way for the creation of innovative healthcare systems. Patients, specifically the elderly who live alone, can be rescued if cloud-based remote patient monitoring models are used. Patients with chronic diseases (e.g. blood pressure problems) could be followed for 24 hours using a cloud-based observing standard, which is valuable in anticipating their health.
- Garg et al., (2019) [41] explained that the volume of organized and amorphous data in the smart city atmosphere has increased dramatically in recent years because of technologies such as edge computing, IoT, and 5G. Therefore, innovative ways of analyzing such large volumes of data are necessary. Surveillance, information and entertainment, and real-time traffic monitoring are just a few of the applications for intelligent transportation systems (ITS), one of the essential components. But the scientific community is worried about its security, performance, and availability. Roadside units (RSUs), cellular networks, and mobile cloud computing (CC) are all currently accessible options that depend on a centralized approach and need the deployment of additional infrastructure, making them far from ideal in their present condition. The suggested framework exceeds previous vehicle representations by delivering a secure, energy-efficient, and minimally delayed approach.
- Shah et al., (2019) [42] stated that when disaster management and the IoT are coupled, a new and more efficient manner of executing fundamental functions could well be realized. It is now feasible not only to gather massive volumes of usable data from various data resources utilizing cutting-edge BD analytics tools and well-managed IoT but also to swiftly generate the insights required to make educated choices. However, given the time restrictions and accuracy requirements of disaster management systems, much more exploration is needed to successfully represent and employ these two concepts. They examined the most current research on disaster management utilizing BDA and IoT in this evaluation of the benefits of these technologies. This brief, which acts as a guide for the project team, also outlines the environment's major demands and concerns. It is concluded that the research might be utilized as a reference to better know the overall functions for making effective use of the BDA and IoT possibilities for disaster management.
- Yassine et al., (2019) [43] stated that IoT analytics are crucial for the development of smart home applications. There is an enormous amount of data collected on the regular routines of consumers by the networked appliances

and devices present in the contemporary smart home. Homeowners, as well as the ever-expanding corporations that want access to customer profiles, could all benefit from IoT data, which can be used to better personalized applications, using cloud and fog nodes to allow data-driven services while meeting the complexity and resource restrictions of online and offline data handling, storage, and classification evaluation. Simulation results clearly illustrate the platform's benefits and use of intended algorithms.

- Zhou et al., (2018) [44] stated that the IoT allows physical objects such as vehicles, appliances, and other devices to communicate and even interact with one another. Smart homes, healthcare, and factory automation all make heavy use of this technology. The IoT has the potential to alter the way one lives and works, but it also offers major security and privacy problems. Even though more research is being conducted to mitigate these risks, many concerns remain unanswered. The investigation first introduces the theory of "IoT characteristics" to recognize the core reasons for growing IoT hazards as well as the limits of present research. Then eight IoT topics are examined, including security and privacy threats, present solutions, and research challenges. This information examines the majority of IoT security analyses from 2013 to 2017 to demonstrate how IoT features affect present security research and to aid academics in staying up to date on the most recent breakthroughs in this sector.
- Vassakis et al., (2018) [45] explained that the change in networks, platforms, and people, as well as the digital technology that drives innovation and competition, all have a significant impact on organizations in the fourth industrial revolution. For businesses, the approaches and techniques used allow them to recognize their environment and make fast decisions on the fly. There has never been a better moment than now to invest in the IoT and BD. Companies appreciate and value the ability to properly handle, assess, and act on their data. BD analytics can boost data-driven organizations' competitiveness and innovative performance. BD could give crucial information, but there are substantial barriers to implementing a data-driven approach and realizing the advantages.
- Habibzadeh et al., (2018) [46] explained that smart parking, smart transportation, and smart environment are just a few of the futuristic applications currently being examined in today's smart city research, which includes an extensive variety of overtly futuristic applications, such as entirely automatic homes and workplaces being able to adjust their temperature to save energy. Developing these functions requires a data collection system that can gather data from many sensors, whether they are linked to permanent infrastructure or integrated into the mobile devices of individuals living in smart cities (soft sensing). Machine information and data analytics are being utilized to give both short-term and long-term benefits in a range of smart city applications. Finally, they discuss some of the outstanding issues and challenges in software-based sensing in smart cities.

- Sun et al., (2016) [47] stated that the needs of remembering the past, living in the present, and planning are all handled synergistically in smart and connected communities (SCC). Therefore, SCC aims to increase a community's livability, conserve it, revive it, and make it accessible. The aim of establishing SCC for a community is to help people remain in the present, prepare for the future, and reflect on the past. It is the progression of the smart city concept into SCC. The past, present, and future are all entwined in SCC, which is why it is being created as a system that operates in tandem with all three (sustainability). The SCC aims to enhance the conditions of life for all citizens to accomplish these goals in a community. Creating SCC allows a community to be present in the now, plan, and remember the past, all at the same time. The IoT and BD analytics can be used to construct a network of connected devices that can be monitored and controlled in real time for SCC. Mobile crowdsensing and CC are two of the most valuable IoT technologies for increasing SCC. The Tree Sight project in Trento, Italy, highlights how IoT and BD analytics can be utilized to promote intelligent tourism and environmental cultural inheritance.
- Mehdipour et al., (2016) [48] explained that as the necessity for real-time processing of massive volumes of data develops, BD analytics is getting increasingly challenging to deploy on present systems. One solution to this challenge is to bring data analytics closer to where the data is produced and stored. The fog-engine technology was created to be integrated into IoTs near the ground and to allow data analysis before large volumes of data are offloaded to a central location. The result shows that it minimizes network bandwidth usage in addition to decreasing latency and enhancing throughput. Table 3.2 depicts the summary of the literature review of statistical and data science challenges and opportunities: a study based on IoT security.

3.3 RESULT AND DISCUSSION

It is the primary premise of this study to compare the accuracy of research on IoT security using ML classifiers, which offer a prediction model based on ML to determine the data analytics based on IoT security. Comparative analysis of studies is performed to identify the performance of ML classifiers by assessing their accuracy. Several authors utilize ML techniques to obtain the accuracy of the system. Comparison among classifiers such as NB, DT, ANN, Ensemble, LR, RF, XGB, AB, SVM, and KNN are performed to determine the accuracy. Author Simran et al. utilizes LR and RF classifiers, which show the highest and lowest accuracy among all classifiers. LR has minimum accuracy rate of 52.6%, and RF has maximum accuracy rate of 99.1%. Table 3.3 depicts the comparative analysis of the accuracy of different classifiers.

Figure 3.4 depicts the comparison graph for the classifiers' accuracy.

TABLE 3.2
Summary of Literature Review

Authors	Techniques	Outcomes
Rehman et al., (2021) [37]	BDA	To develop the foundation for BD analytics in healthcare, government agencies, healthcare professionals, hardware manufacturers, and pharmaceutical organizations; people, data scientists, academia, and suppliers must cooperate.
Kumari et al., (2020) [38]	ICT	Secure analytics from BD have proven to be game changers in a multitude of smart city applications, including SG and transportation.
Bhattarai et al., (2019) [39]	Power grid technology	According to an issue of the *Journal of Business Analytics*, utility firms could establish new business standards and profit flows by utilizing BD analytics in several ways. Consequently, the utility companies will be able to make the best financing and functioning outcomes at the right point and in the right location.
Hassan et al., (2019) [40]	RPM	Patients with chronic diseases (e.g. blood pressure problems) could be followed for 24 hours using a cloud-based monitoring simulation, which is valuable in anticipating their health.
Garg et al., (2019) [41]	ITS	The suggested framework exceeds previous vehicle representations by delivering a secure, energy-efficient, and minimally delayed approach.
Shah et al., (2019) [42]	BDA	It is concluded that the research might be utilized as a reference to better know the overall functions for making effective use of the BDA and IoT possibilities for disaster management.
Yassine et al., (2019) [43]	Fog	Simulation results clearly illustrate the platform's benefits and use of intended algorithms.
Zhou et al., (2018) [44]	IoT	This information examines the majority of IoT security analyses from 2013 to 2017 to demonstrate how IoT features affect present security research and to aid academics in staying up to date on the most recent breakthroughs in this sector.
Vassakis et al., (2018) [45]	BDA	BD could provide crucial information, but there are substantial barriers to implementing a data-driven approach and realizing the advantages.
Habibzadeh et al., (2018) [46]	IoT	Machine intelligence and data analytics are being utilized to give both short-term and long-term benefits in a range of smart city applications.
Sun et al., (2016) [47]	CC	The Tree Sight project in Trento, Italy, highlights how IoT and BD analytics can be utilized to promote smart tourism and sustainable cultural heritage.
Mehdipour et al., (2016) [48]	Fog-engine technology	The result shows that it minimizes network bandwidth usage in addition to decreasing latency and enhancing throughput.

TABLE 3.3
Comparison of Accuracy of Different Classifiers

Authors	Techniques	Accuracy (%)
Hassan et al. [40]	NB	87.6
	DT	83.8
Moustafa et al. [49]	ANN	94.22
	Ensemble	98.29
Sriram et al. [50]	LR	52.6
	RF	99.1
Verma et al. [51]	XGB	93.15
	AB	90.37
Qureshi et al. [52]	SVM	77.5
	KNN	76

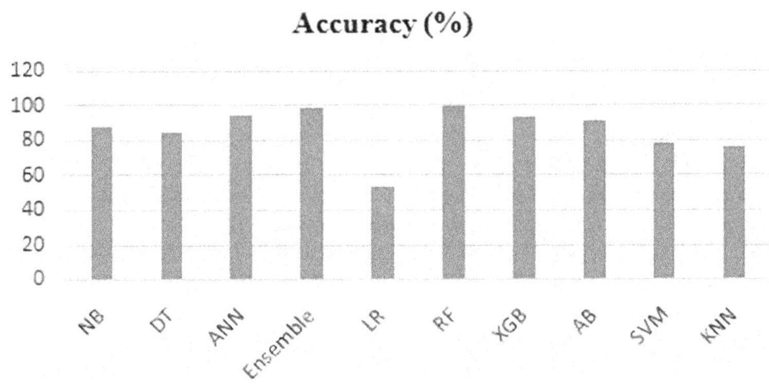

FIGURE 3.4 Graphical representation of the accuracy of various classifiers.

3.4 CONCLUSION

The emergence of smart sensor devices has resulted in a remarkable increase in data generated during the past several years. When engaging with IoT and BD, it is now critical to handle, transform, and analyze the data quickly. This survey was conducted as part of a large-scale IoT data study. IoT and BD analytics were also investigated. A structural design for big IoT data analytics is provided. Additionally, large data analytics forms, procedures, and technologies for BD extracting were examined. There were also real-world applications. As part of the examination of the IoT space, it addresses many possibilities that data analytics offers. As a result, the essential interventions can be carried out at the most opportune point in the treatment process.

The accuracy rate and classifiers employed in several prior types of studies for statistical data science challenges based on IoT security in the examination, and the findings of those studies are utilized to compare the results of this study. This research is based on a comparative analysis of classifiers that are used by various authors for the statistical data science challenges based on IoT security. The random forest classifier has the highest accuracy rate of 99.1%. Properly examining the effect of new elements on security and privacy in the IoT can help better forecast future research hotspots and the evolution of IoT security. Future research directions were discussed regarding several unanswered research topics.

REFERENCES

[1] Al Nuaimi, E., Al Neyadi, H., Mohamed, N., & Al-Jaroodi, J. (2015). Applications of big data to smart cities. *Journal of Internet Services and Applications*, *6*(1), 1–15.

[2] Gubbi, J., Buyya, R., Marusic, S., & Palaniswami, M. (2013). Internet of things (IoT): A vision, architectural elements, and future directions. *Future Generation Computer Systems*, *29*(7), 1645–1660.

[3] Hsieh, H. C., & Lai, C. H. (2011, December). Internet of things architecture based on integrated PLC and 3G communication networks. In *2011 IEEE 17th International Conference on Parallel and Distributed Systems* (pp. 853–856). IEEE.

[4] Kambatla, K., Kollias, G., Kumar, V., & Grama, A. (2014). Trends in big data analytics. *Journal of Parallel and Distributed Computing*, *74*(7), 2561–2573.

[5] Hashem, I. A. T., Yaqoob, I., Anuar, N. B., Mokhtar, S., Gani, A., & Khan, S. U. (2015). The rise of "big data" on cloud computing: Review and open research issues. *Information Systems*, *47*, 98–115.

[6] Ali, W. B. (2016). Big data-driven smart policing: Big data-based patrol car dispatching. *Journal of Geotechnical and Transportation Engineering*, *1*(2), 1.

[7] Gantz, J., & Reinsel, D. (2012). The digital universe in 2020: Big data, bigger digital shadows, and biggest growth in the far east. *IDC iView: IDC Analyze the Future*, *2007*(2012), 1–16.

[8] Marjani, M., Nasaruddin, F., Gani, A., Karim, A., Hashem, I. A. T., Siddiqa, A., & Yaqoob, I. (2017). Big IoT data analytics: Architecture, opportunities, and open research challenges. *IEEE Access*, *5*, 5247–5261.

[9] Turner, V., Gantz, J. F., Reinsel, D., & Minton, S. (2014). The digital universe of opportunities: Rich data and the increasing value of the internet of things. *IDC Analyze the Future*, *16*, 13–19.

[10] Lee, Y. T., Hsiao, W. H., Lin, Y. S., & Chou, S. C. T. (2017). Privacy-preserving data analytics in cloud-based smart home with community hierarchy. *IEEE Transactions on Consumer Electronics*, *63*(2), 200–207.

[11] Iqbal, A., Suryani, M. A., Saleem, R., & Suryani, M. A. (2016). Internet of things (IoT): On-going security challenges and risks. *International Journal of Computer Science and Information Security*, *14*(11), 671.

[12] Celik, Z. B., Babun, L., Sikder, A. K., Aksu, H., Tan, G., McDaniel, P., & Uluagac, A. S. (2018). Sensitive information tracking in commodity IoT. In *Proceedings of the 27th USENIX Security Symposium* (pp. 1687–1704). USENIX Association.

[13] Leverett, É., Clayton, R., & Anderson, R. (2017). Standardisation and certification of the 'internet of things'. *Paper presented at Workshop on the Economics of Information Security 2017*. San Diego, CA, USA.

[14] Celik, Z. B., Fernandes, E., Pauley, E., Tan, G., & McDaniel, P. (2019). Program analysis of commodity IoT applications for security and privacy: Challenges and opportunities. *ACM Computing Surveys (CSUR)*, *52*(4), 1–30.

[15] Kubler, S., Främling, K., & Buda, A. (2015). A standardized approach to deal with firewall and mobility policies in the IoT. *Pervasive and Mobile Computing*, *20*, 100–114.
[16] He, W, Golla, M., Padhi, R., Ofek, J., Dürmuth, M., Fernandes, E., & Ur, B. (2018). Rethinking access control and authentication for the home internet of things (IoT). In *Proceedings of the 27th USENIX Conference on Security Symposium (SEC'18)* (pp. 255–272). USENIX Association, USA.
[17] Leiba, O., Yitzchak, Y., Bitton, R., Nadler, A., & Shabtai, A. (2018, April). Incentivized delivery network of IoT software updates based on trustless proof-of-distribution. In *2018 IEEE European Symposium on Security and Privacy Workshops (EuroS&PW)* (pp. 29–39). IEEE.
[18] Tian, Y., Zhang, N., Lin, Y. H., Wang, X., Ur, B., Guo, X., & Tague, P. (2017). {SmartAuth}: {User-Centered} authorization for the internet of things. In *26th USENIX Security Symposium (USENIX Security 17)* (pp. 361–378).
[19] Tawalbeh, L. A., Muheidat, F., Tawalbeh, M., & Quwaider, M. (2020). IoT privacy and security: Challenges and solutions. *Applied Sciences*, *10*(12), 4102.
[20] Mishra, S., & Tripathi, A. R. (2020). IoT platform business model for innovative management systems. *International Journal of Financial Engineering*, *7*(03), 2050030.
[21] Mishra, S., & Tripathi, A. R. (2020). Literature review on business prototypes for digital platform. *Journal of Innovation and Entrepreneurship*, *9*(1), 1–19.
[22] Mishra, S., & Tripathi, A. R. (2021). AI business model: An integrative business approach. *Journal of Innovation and Entrepreneurship*, *10*(1), 18.
[23] Mishra, S., & Mishra, P. (2022). Analysis of platform business and secure business intelligence. *International Journal of Financial Engineering*, *9*(03), 2250002.
[24] Mohideen, B. I., & Assiri, B. (2021). Internet of things (IoT): Classification, secured architecture based on data sensitivity, security issues and their countermeasures. *Journal of Information & Knowledge Management*, *20*(supp01), 2140001.
[25] Sun, L., & Li, M. (2021). Sports and health management using big data based on voice feature processing and internet of things. *Scientific Programming*, *2021*.
[26] Fatima, H., Khan, H. U., & Akbar, S. (2021). Home automation and RFID-based internet of things security: Challenges and issues. *Security and Communication Networks*, *2021*, 1–21.
[27] Zaldivar, D., Lo'Ai, A. T., & Muheidat, F. (2020, January). Investigating the security threats on networked medical devices. In *2020 10th Annual Computing and Communication Workshop and Conference (CCWC)* (pp. 0488–0493). IEEE.
[28] Liu, X., Zhao, M., Li, S., Zhang, F., & Trappe, W. (2017). A security framework for the internet of things in the future internet architecture. *Future Internet*, *9*(3), 27.
[29] Lo'Ai, A. T., & Somani, T. F. (2016, November). More secure Internet of things using robust encryption algorithms against side channel attacks. In *2016 IEEE/ACS 13th International Conference of Computer Systems and Applications (AICCSA)* (pp. 1–6). IEEE.
[30] El-Hajj, M., Fadlallah, A., Chamoun, M., & Serhrouchni, A. (2019). A survey of internet of things (IoT) authentication schemes. *Sensors*, *19*(5), 1141.
[31] Iqbal, W., Abbas, H., Daneshmand, M., Rauf, B., & Bangash, Y. A. (2020). An in-depth analysis of IoT security requirements, challenges, and their countermeasures via software-defined security. *IEEE Internet of Things Journal*, *7*(10), 10250–10276.
[32] Tanwar, S., Tyagi, S., & Kumar, S. (2018). The role of internet of things and smart grid for the development of a smart city. In *Intelligent Communication and Computational Technologies: Proceedings of Internet of Things for Technological Development, IoT4TD 2017* (pp. 23–33). Springer Singapore.
[33] Bhattarai, B. P., Paudyal, S., Luo, Y., Mohanpurkar, M., Cheung, K., Tonkoski, R., . . . Zhang, X. (2019). Big data analytics in smart grids: State-of-the-art, challenges, opportunities, and future directions. *IET Smart Grid*, *2*(2), 141–154.
[34] Munir, M., Baumbach, S., Gu, Y., Dengel, A., & Ahmed, S. (2018). Data analytics: Industrial perspective & solutions for streaming data. In *Series in Machine Perception and Artificial Intelligence* (pp. 144–168). doi: 10.1142/9789813228047_0007.

[35] Krommyda, M., & Kantere, V. (2021). Spatial data management in IoT systems: Solutions and evaluation. *International Journal of Semantic Computing*, *15*(01), 117–139.

[36] Balashunmugaraja, B., & Ganeshbabu, T. R. (2020). Optimal key generation for data sanitization and restoration of cloud data: Future of financial cyber security. *International Journal of Information Technology & Decision Making*, *19*(04), 987–1013.

[37] Rehman, A., Naz, S., & Razzak, I. (2022). Leveraging big data analytics in healthcare enhancement: Trends, challenges and opportunities. *Multimedia Systems*, *28*(4), 1339–1371.

[38] Kumari, A., & Tanwar, S. (2020). Secure data analytics for smart grid systems in a sustainable smart city: Challenges, solutions, and future directions. *Sustainable Computing: Informatics and Systems*, *28*, 100427.

[39] Bhattarai, B. P., Paudyal, S., Luo, Y., Mohanpurkar, M., Cheung, K., Tonkoski, R., . . . Zhang, X. (2019). Big data analytics in smart grids: State-of-the-art, challenges, opportunities, and future directions. *IET Smart Grid*, *2*(2), 141–154.

[40] Hassan, M. K., El Desouky, A. I., Elghamrawy, S. M., & Sarhan, A. M. (2019). Big data challenges and opportunities in healthcare informatics and smart hospitals. In Hassanien, A., Elhoseny, M., Ahmed, S., & Singh, A. (eds) *Security in Smart Cities: Models, Applications, and Challenges. Lecture Notes in Intelligent Transportation and Infrastructure*. Springer, Cham. doi: 10.1007/978-3-030-01560-2_1.

[41] Garg, S., Singh, A., Kaur, K., Aujla, G. S., Batra, S., Kumar, N., & Obaidat, M. S. (2019). Edge computing-based security framework for big data analytics in VANETs. *IEEE Network*, *33*(2), 72–81.

[42] Shah, S. A., Seker, D. Z., Hameed, S., & Draheim, D. (2019). The rising role of big data analytics and IoT in disaster management: Recent advances, taxonomy and prospects. *IEEE Access*, *7*, 54595–54614.

[43] Yassine, A., Singh, S., Hossain, M. S., & Muhammad, G. (2019). IoT big data analytics for smart homes with fog and cloud computing. *Future Generation Computer Systems*, *91*, 563–573.

[44] Zhou, W., Jia, Y., Peng, A., Zhang, Y., & Liu, P. (2018). The effect of IoT new features on security and privacy: New threats, existing solutions, and challenges yet to be solved. *IEEE Internet of Things Journal*, *6*(2), 1606–1616.

[45] Vassakis, K., Petrakis, E., & Kopanakis, I. (2018). Big data analytics: Applications, prospects and challenges. In Skourletopoulos, G., Mastorakis, G., Mavromoustakis, C., Dobre, C., & Pallis, E. (eds) *Mobile Big Data. Lecture Notes on Data Engineering and Communications Technologies, vol 10*. Springer, Cham. doi: 10.1007/978-3-319-67925-9_1.

[46] Habibzadeh, H., Boggio-Dandry, A., Qin, Z., Soyata, T., Kantarci, B., & Mouftah, H. T. (2018). Soft sensing in smart cities: Handling 3Vs using recommender systems, machine intelligence, and data analytics. *IEEE Communications Magazine*, *56*(2), 78–86.

[47] Sun, Y., Song, H., Jara, A. J., & Bie, R. (2016). Internet of things and big data analytics for smart and connected communities. *IEEE Access*, *4*, 766–773.

[48] Mehdipour, F., Javadi, B., & Mahanti, A. (2016, August). FOG-engine: Towards big data analytics in the fog. In *2016 IEEE 14th International Conference on Dependable, Autonomic and Secure Computing, 14th International Conference on Pervasive Intelligence and Computing, 2nd International Conference on Big Data Intelligence and Computing and Cyber Science and Technology Congress (DASC/PiCom/DataCom/CyberSciTech)* (pp. 640–646). IEEE.

[49] Moustafa, N., Turnbull, B., & Choo, K. K. R. (2018). An ensemble intrusion detection technique based on proposed statistical flow features for protecting network traffic of internet of things. *IEEE Internet of Things Journal*, *6*(3), 4815–4830.

[50] Sriram, S., Vinayakumar, R., Alazab, M., & Soman, K. P. (2020, July). Network flow based IoT botnet attack detection using deep learning. In *IEEE INFOCOM 2020-IEEE Conference on Computer Communications Workshops (INFOCOM WKSHPS)* (pp. 189–194). IEEE.
[51] Verma, A., & Ranga, V. (2020). Machine learning based intrusion detection systems for IoT applications. *Wireless Personal Communications*, *111*, 2287–2310.
[52] Qureshi, K. N., Din, S., Jeon, G., & Piccialli, F. (2020). An accurate and dynamic predictive model for a smart M-Health system using machine learning. *Information Sciences*, *538*, 486–502.

4 Advanced Machine Learning Strategies for Road Object Detection in Multimedia Environments

Preeti, Chhavi Rana

4.1 INTRODUCTION

The development of autonomous vehicles has gained significant momentum in recent years, and it is anticipated to revolutionize the transportation industry [1]. Autonomous vehicles rely on a multitude of sensors, cameras, and other hardware to collect information about their surroundings, which is then processed and analyzed by artificial intelligence (AI) algorithms. One of the critical functions of autonomous vehicles is road object detection, which involves identifying and tracking objects on the road, such as other vehicles, pedestrians, and obstacles.

AI-based road object detection has emerged as a key technology for autonomous vehicles. Deep learning–based object detection algorithms have achieved remarkable accuracy in detecting objects on the road [2]. However, these algorithms are often perceived as black boxes, making it challenging to understand their decision-making process. This lack of interpretability and transparency in AI-based road object detection has raised concerns about safety, accountability, and ethical issues.

In order to address these concerns, researchers are developing explanation techniques for AI-based road object detection, which provide insights into how the algorithms make decisions. Explanation techniques offer a way to understand the reasoning behind the object detection algorithms, allowing developers to identify and address any biases or errors in the system. As a result, users and officials will have an easier time determining the security and dependability of autonomous vehicles if explanation approaches are implemented.

This study provides a comprehensive review of AI-based road object detection through explanations. This chapter discusses the different techniques used for AI-based road object detection, including traditional and deep learning–based approaches. The importance of interpretability in AI-based road object detection is highlighted, and the performance comparison of different techniques is discussed. The study also explains the importance of road object detection for autonomous

 DOI: 10.1201/9781003477280-4

vehicles and identifies the various approaches used in road object detection. Finally, the chapter presents the future of AI-based road object detection, including advancements in the technology, its integration into autonomous vehicles, and its potential impact on the transportation industry.

4.1.1 Machine Learning–Based Road Object Detection

AI-based road object detection is the process of identifying and tracking objects on the road using artificial intelligence algorithms. The objects detected can include other vehicles, pedestrians, cyclists, and obstacles such as debris, potholes, and road signs. Road object detection is an essential function of autonomous vehicles and is critical for ensuring their safety and reliability [3]. Figure 4.1 illustrates AI-based real-time object detection.

Traditional approaches to road object detection involve handcrafting features and using machine learning algorithms to classify objects based on those features. However, these methods require a significant amount of domain knowledge, are time consuming, and are often prone to errors. With the advent of deep learning algorithms, object detection has been revolutionized, allowing for faster and more accurate detection [5].

4.1.2 Various Approaches Used in Road Object Detection

Autonomous vehicles rely heavily on object detection technology, and recent years have seen a rise in the popularity of deep learning–based object detection algorithms [6]. Convolutional neural networks (CNNs) are used in these algorithms because they are optimal for picture recognition. In order to categorize the objects in an image, they first use a series of convolutional layers to extract information from the image.

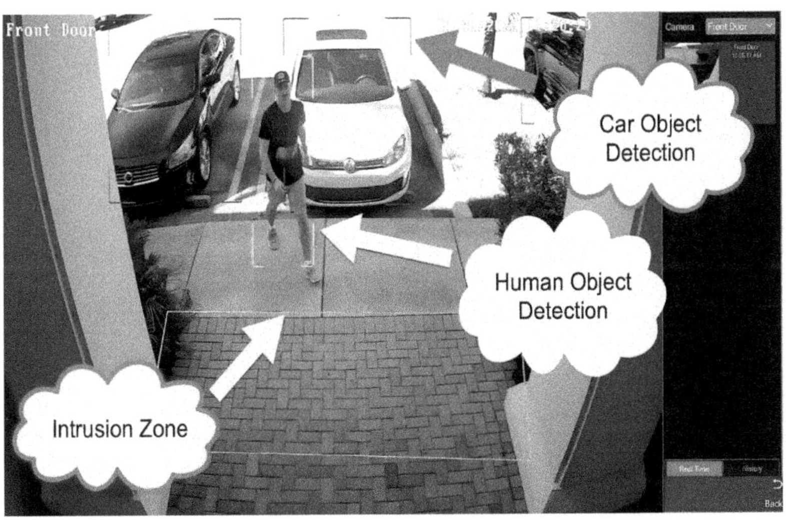

FIGURE 4.1 Machine learning–based object detection [4].

One of the most widely used deep learning–based object detection algorithms, R-CNN, uses a selective search technique to provide region recommendations that are highly likely to include an item [7]. After these suggestions have been fed into a convolutional neural network, features can be extracted and used for object classification and further refinement of the region proposal.

Faster R-CNN, SSD, and YOLO are just a few of the deep learning–based object identification algorithms that have increased the precision and throughput of autonomous vehicle road object recognition. Figure 4.2 describes autonomous vehicle scene perception and object detection using deep learning architectures.

However, there are concerns about their lack of interpretability and transparency. A common misconception about deep learning algorithms is that their decision-making processes are opaque. This lack of transparency can lead to issues with bias, accountability, and safety. To address these concerns, researchers are developing explanation techniques for AI-based road object detection. Explanation techniques aim to provide insights into how the algorithms make decisions, allowing developers to identify and address any biases or errors in the system. Examples of explanation techniques include feature visualization, saliency maps, and attention maps. These techniques are crucial for ensuring the safe and responsible development of autonomous vehicles and their widespread adoption in the future.

4.1.3 Importance of Road Object Detection for Autonomous Vehicles

Road object detection is a crucial technology for autonomous vehicles. It allows vehicles to perceive and interpret their environment, identify potential hazards, and make

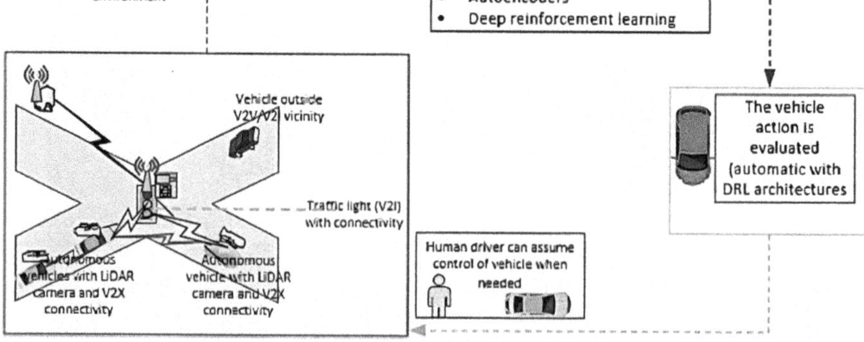

FIGURE 4.2 Autonomous vehicle scene perception and object detection using deep learning architectures [8].

decisions in real time to avoid collisions and ensure the safety of passengers and other road users. Here are some reasons road object detection is essential for autonomous vehicles [9, 10]:

1. **Safety**: Road object detection is critical for ensuring the safety of autonomous vehicles. Avoiding and preventing accidents is made possible by the vehicle's ability to detect and identify items on the road, such as other vehicles, pedestrians, and obstructions.
2. **Navigation**: Autonomous vehicles rely on road object detection to navigate their surroundings. By detecting road signs, lane markers, and traffic lights, the vehicle can accurately position itself and follow the correct path.
3. **Efficiency**: Road object detection can improve the efficiency of autonomous vehicles by allowing them to optimize their driving patterns. For example, by detecting the speed and direction of other vehicles, the vehicle can adjust its speed and lane position to reduce congestion and improve traffic flow.
4. **Reliability**: Autonomous vehicles need to be reliable and consistent in their behavior to gain public trust and acceptance. Road object detection is critical for ensuring the reliability of autonomous vehicles by providing accurate and consistent information about the vehicle's surroundings.
5. **Adaptability**: Road object detection can help autonomous vehicles adapt to changing road conditions, such as weather, road construction, and traffic patterns. By detecting changes in the environment, the vehicle can adjust its behaviour to ensure safe and efficient driving.

In summary, road object detection is essential for the development and implementation of autonomous vehicles. It enables vehicles to perceive and interpret their environment, navigate their surroundings, and ensure the safety and reliability of passengers and other road users. Without road object detection, autonomous vehicles would not be able to operate safely and efficiently in complex and dynamic environments.

4.2 LITERATURE OF REVIEW

Several methods for detecting objects on the road ahead of autonomous vehicles presented by various authors are given in this section.

- Mankodiya et al. (2022) [11] analyzed three semantic segmentation architectures and compared them to deep learning (DL) models for pixel-wise road detection. DL algorithms are often considered complex, making it difficult to interpret their highly complex structures. In order to analyze and comprehend the models' predictions, the study implemented explainable artificial intelligence (XAI) methods. This study used several XAI techniques to generate rationales for the segmentation model presented for AV road detection. Each model's efficacy was visualized by comparing its outputs to ground-truth images. When compared to ResNet-50 and SegNet, which achieved accuracies of 0.9281 and 0.9655 respectively, ResNet-18 accomplished the highest accuracy of 0.9761 while generating sharper boundaries on road edges.

- Wang et al. (2021) [12] discussed the importance of real-time and accurate object detection in autonomous driving technology. The study presented herein established a YOLOv4-based, single-stage object detection method with enhanced detection accuracy and real-time functionality. In order to accomplish this, the algorithm enhances the YOLOv4's spine, neck, and feature fusion module. The study reports a 2.06% improvement in average accuracy on the KITTI dataset and 2.95% on the BDD dataset compared to YOLOv4 while also increasing inference speed by 9.14%. The authors achieved this by adding attention mechanisms and reducing the overall width of the network to reduce model parameters. Overall, the proposed algorithm strikes a balance between speed and accuracy for autonomous driving.
- Ponn et al. (2020) [13] emphasized the importance of identifying and evaluating risks associated with automated vehicles before their market launch. The purpose of this study was to offer a modeling strategy for object detection utilizing cameras and AI algorithms. The detection performance of numerous object identification algorithms was analyzed and explained using the SHapley Additive exPlanations (SHAP) method. Their research showed that several influencing elements, such as an object's relative rotation towards the camera or its position on the image, affect the detection performance in the same way, independent of the detection technique. The research showed that random forests could simulate object detectors' detection performances with an accuracy of up to 84.8%, which is more than adequate for practical use. Overall, the study provides insights into the importance of evaluating risks associated with automated vehicles and the effectiveness of modeling approaches in analyzing object detection algorithms.
- Nabati et al. (2019) [14] emphasized the use of region proposal algorithms in two-phase object detection networks for predicting the location of objects in an image. Real-time applications, such as autonomous driving, can't handle the additional processing time introduced by region proposal algorithms. The authors of this study proposed a radar-based real-time region proposal method (RRPN) that does this by translating radar detections into the image's coordinate system and then creating anchor boxes for each one. In order to make more precise suggestions, these boxes are altered and scaled according to the object's distance from the vehicle. On the NuScenes dataset, the proposed model was found to be faster by a factor of over 100, compared to the Selective Search object proposal algorithm, all while improving detection precision and recall.
- He et al. (2017) [15] developed an effective method for detecting objects in images and producing a high-quality segmentation mask for each instance using a generic and adaptable framework they named Mask R-CNN. This technique expands Faster R-CNN by incorporating a second branch, in addition to the original's bounding box recognition functionality, to predict a mask for an item. In addition to being fast (5 fps) and easy to train, Mask

R-CNN may be applied to a variety of tasks, including human posture estimation. The proposed technique performed exceptionally well across all three domains of the COCO set of challenges. Mask R-CNN easily bested the COCO 2016 challenge champions and every other single-model entry across the board. The authors expect this straightforward method will pave the way for additional studies on instance-level recognition.

- Lin et al. (2017) [16] proposed a method called RetinaNet for object detection which uses a one-stage approach and achieves high accuracy while maintaining speed. Traditional one-stage detectors have not performed as well as two-stage detectors, but this study found that the reason for this was excessive main image-background class disparity during training. Focal Loss, developed by the authors to solve this problem, reduces the weight given to the loss associated with correctly categorized examples and instead directs attention to training on a small subset of challenging examples. The RetinaNet detector was trained using this loss function and achieved state-of-the-art accuracy. The findings of this study demonstrate the potential for improving the accuracy of one-stage detectors and their practicality for real-time applications.
- Dai et al. (2016) [17] recommended a novel strategy for detecting objects by utilizing fully convolutional networks that are region based. Their method is fully convolutional, with most of the computation being shared across the entire image, as opposed to previous methods that required a per-region subnetwork to be applied hundreds of times. They introduce position-sensitive score maps to resolve the conflict between translation invariance in picture classification and translation variability in object detection. As a result, they may perform object detection using fully convolutional image classifier backbones, such as Residual Networks. Using the 101-layer ResNet, they produce a competitive performance on the PASCAL VOC datasets. Their mAP for the 2007 set is 83.6%. The test-time speed of their method is 170 ms per image, which is 2.5–20 faster than the Faster R-CNN analog. In general, the fully convolutional network method outperforms other approaches in terms of accuracy and speed when it comes to detecting objects.
- Liu et al. (2016) [18] introduced the use of a single deep neural network, named SSD, to do picture object recognition. This method employs a discrete set of default boxes with varying aspect ratios and scales for each feature map point to generate bounding boxes. At the time of prediction, the network assigns weights to each possible category of objects within each predefined box and refines the box's shape accordingly. To deal with objects of varying sizes, several feature maps are integrated. As shown by experiments on many datasets, SSD is as accurate as approaches that employ object proposal phases while being significantly faster to process. SSD outperformed state-of-the-art models on the VOC2007 test, achieving 74.3% mAP1 for 300 300 input at 59 fps on an Nvidia Titan X and 76.9% mAP for 512 512 input. When compared to other single-stage methods, SSD's accuracy improves even when the input image size is reduced.

- Uçar et al. (2016) [19] explored the significance of object recognition and pedestrian detection in autonomous driving applications. The study proposes two CNN architectures with various layers and extracts features from these networks using Speeded Up Robust Feature (SURF), Histogram of Gradient (HOG), and k-means. The study also uses the Bag of Visual Words (BOW) approach to extract features. Linear Support Vector Machine (SVM) classifiers are used to train the features, and experiments are conducted using Caltech 101 and Caltech Pedestrian Detection datasets. The study's findings indicate that deep learning–based approaches with Compute Unified Device Architecture (CUDA) support provide more accurate performance as well as quicker decision-making in applications that run in real time.
- Girshick et al. (2014) [20] revealed that advancements in object detection performance on the PASCAL VOC dataset have stalled. The most successful approaches involved intricate ensemble systems that integrated numerous low-level visual elements with high-level context. In order to solve this problem, they created an easy-to-implement detection algorithm dubbed R-CNN, which increased mean average precision by almost 30% compared to earlier approaches. The R-CNN method for object localization and segmentation utilizes a mixture of high-capacity CNNs and bottom-up region recommendations. They propose employing supervised pretraining for an ancillary task, followed by domain-particular fine-tuning to increase performance in situations in which labelled training data is insufficient. The authors also presented experiments that revealed the network's hierarchy of image features.

The methods and results of the reviewed literature presented by various authors are summarized in Table 4.1:

4.3 COMPARATIVE ANALYSIS

This section provides a comparative analysis of the performance of several AI-based road object detection models for autonomous vehicles. The models considered in the analysis are ResNet-18, ResNet-50, SegNet, YOLOv4, Mask R-CNN, R-FCN, RetinaNet, and SSD.

Models used for road object recognition in autonomous vehicles are compared and contrasted in Table 4.2. The highest accuracy was achieved by ResNet-18 with an accuracy of 97.61%, followed by SegNet with an accuracy of 96.55%. These two models have fewer layers than other models but have shown superior performance. The ResNet-50 model also performed well with an accuracy of 92.81% but not as well as ResNet-18. On the other hand, YOLOv4 showed an accuracy of 88.67%, which is significantly lower than the ResNet models. However, YOLOv4 has the fewest layers among the models considered. This suggests that YOLOv4 may be a suitable option for autonomous vehicles that require faster processing times.

Mask R-CNN, RetinaNet, R-FCN, and SSD showed relatively lower accuracy levels of 67.61%, 62.57%, 52.86%, and 41.52% respectively. These models have more layers than the other models but have not shown superior performance, as observed

TABLE 4.1
Comparison of Reviewed Literature

Authors	Techniques	Outcomes
Mankodiya et al. (2022) [11]	ResNet-50 ResNet-18 SegNet	Results showed that ResNet-18 achieved the highest accuracy of 0.9761; ResNet-50 and SegNet had accuracies of 0.9281 and 0.9655 respectively.
Wang et al. (2021) [12]	YOLOv4	The study reports a 2.06% improvement in average accuracy on the KITTI dataset and 2.95% on the BDD dataset compared to YOLOv4 while also increasing inference speed by 9.14%.
Ponn et al. (2020) [13]	SHAP Random forests	The study revealed that random forests could simulate object detectors' detection performances with an accuracy of up to 84.8%, which is more than adequate for practical use.
Nabati et al, (2019) [14]	RRPN	The proposed model was evaluated on the NuScenes dataset, and it was found that RRPN was over 100 times faster and achieved higher detection precision and recall.
He et al. (2017) [15]	Mask R-CNN	The results revealed that Mask R-CNN easily bested the COCO 2016 challenge champions and every other single-model entry across the board.
Lin et al. (2017) [16]	RetinaNet	The findings of this study demonstrate the potential for improving the accuracy of one-stage detectors and their practicality for real-time applications.
Dai et al. (2016) [17]	R-FCN	The proposed method achieves competitive results with an mAP of 83.6% using the 101-layer ResNet, and the approach is found to be considerably faster than the Faster R-CNN.
Liu et al. (2016) [18]	SSD	According to the findings, the SSD has achieved 76.9% accuracy, which is higher than any competing state-of-the-art model.
Uçar et al. (2016) [19]	CNNs BOW	The results of the study suggest that deep learning–based methods supported by CUDA offer improved accuracy and faster decision-making in real-time applications.
Girshick et al. (2014) [20]	R-CNN	The proposed R-CNN technique improved the mean average precision by over 30% compared to previous methods.

in Table 4.2. The lower accuracy levels may be due to overfitting as the models may have been trained with large amounts of data and complex features that may not be relevant to the road object detection problem.

Overall, Table 4.2 provides insights into the performance of several AI-based road object detection models for autonomous vehicles. The analysis suggests that ResNet-18, SegNet, and ResNet-50 are suitable options for achieving high accuracy

TABLE 4.2
Comparison of Different Models

Authors	Models	Layers	Accuracy
Mankodiya et al. (2022) [11]	ResNet-18	111	97.61
	ResNet-50	215	92.81
	SegNet	65	0.9655
Wang et al. (2021) [12]	YOLOv4	53	88.67
He et al. (2017) [15]	Mask R-CNN	101	67.61
Lin et al. (2017) [16]	RetinaNet	101	62.57
Dai et al. (2016) [17]	R-FCN	101	52.86
Liu et al. (2016) [18]	SSD	16	41.52

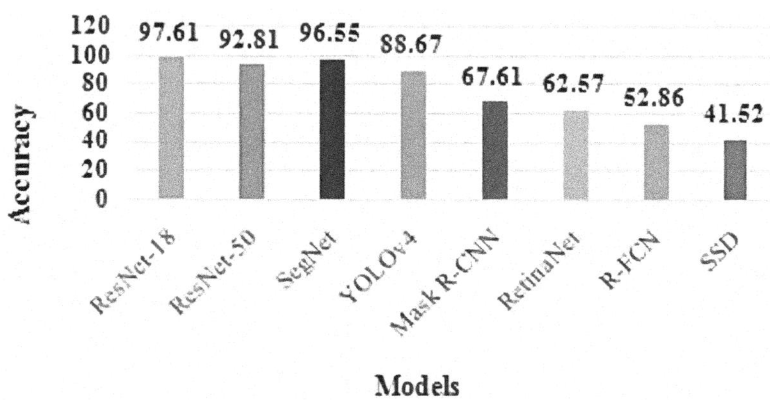

FIGURE 4.3 Comparison of different models used for road object detection models for autonomous vehicles.

levels while YOLOv4 may be a suitable option for achieving faster processing times. However, the selection of a model should be based on the specific requirements of the autonomous vehicle and the environment it will be operating in. The graphical representation of the accuracy of each model is illustrated in Figure 4.3.

4.4 CONCLUSION

The development of AI-based road object detection models for autonomous vehicles is crucial for the safe and efficient operation of these vehicles. The comparative analysis presented in this study provides valuable insights into the performance of several models and can aid in the selection of appropriate models for specific use cases. The high accuracy levels achieved by ResNet-18, SegNet, and ResNet-50 highlight the effectiveness of these models in accurately detecting road objects. These models have a relatively lower number of layers, which may make them computationally less expensive while still providing superior performance. It's worth noting, though, that many variables can affect a model's precision, such as the size and quality of the training dataset, the features used for training, and the complexity of the objects to be detected. The lower accuracy levels achieved by Mask R-CNN, RetinaNet, R-FCN, and SSD suggest that these models may not be as effective in detecting road objects as the other models. However, these models may still be suitable for certain use cases, especially if faster processing times are a priority. The results of the analysis suggest that YOLOv4 may be a suitable option for autonomous vehicles that require faster processing times. This model has the fewest layers of the models considered, which makes it computationally efficient. However, the lower accuracy level of YOLOv4 suggests it may not be as effective in detecting road objects as the other models.

Future research in AI-based road object detection can focus on improving the accuracy levels of the models by training them with more diverse and larger datasets. Furthermore, new models can be developed that integrate other sensor data, such as LiDAR and radar, to enhance the detection and classification of objects. Additionally, the development of explainable AI-based models can help increase trust and transparency in the decision-making process of autonomous vehicles. Finally, the integration of AI-based road object detection with other technologies, such as edge computing and 5G, can enable faster and more efficient processing of the data in real time.

REFERENCES

[1] Ma, Yifang, Zhenyu Wang, Hong Yang, and Lin Yang. "Artificial intelligence applications in the development of autonomous vehicles: A survey." *IEEE/CAA Journal of Automatica Sinica* 7, no. 2 (2020): 315–329.

[2] Ahangar, M. Nadeem, Qasim Z. Ahmed, Fahd A. Khan, and Maryam Hafeez. "A survey of autonomous vehicles: Enabling communication technologies and challenges." *Sensors* 21, no. 3 (2021): 706.

[3] Girshick, R., J. Donahue, T. Darrell, and J. Malik. Rich feature hierarchies for accurate object detection and semantic segmentation. arXiv 2014, arXiv:1311.2524.

[4] https://videos.cctvcamerapros.com/v/ai-security-camera-object-detection.html.

[5] Kim, Keonhyeong. "Secure object detection based on deep learning." *Journal of Information Processing Systems* 17, no. 3 (2021): 571–585.

[6] Saranya, M., N. Archana, J. Reshma, S. Sangeetha, and M. Varalakshmi. "Object detection and lane changing for self-driving car using CNN." In *2022 International Conference on Communication, Computing and Internet of Things (IC3IoT)*, pp. 1–7. IEEE (2022).

[7] Haris, Malik, and Adam Glowacz. "Road object detection: A comparative study of deep learning-based algorithms." *Electronics* 10, no. 16 (2021): 1932.

[8] Gupta, Abhishek, Alagan Anpalagan, Ling Guan, and Ahmed Shaharyar Khwaja. "Deep learning for object detection and scene perception in self-driving cars: Survey, challenges, and open issues." *Array* 10 (2021): 100057.

[9] Chen, Xiaozhi, Kaustav Kundu, Yukun Zhu, Andrew G. Berneshawi, Huimin Ma, Sanja Fidler, and Raquel Urtasun. "3D object proposals for accurate object class detection." *Advances in Neural Information Processing Systems* 28 (2015).

[10] Mahaur, Bharat, Navjot Singh, and K. K. Mishra. "Road object detection: A comparative study of deep learning-based algorithms." *Multimedia Tools and Applications* 81, no. 10 (2022): 14247–14282.

[11] Mankodiya, Harsh, Dhairya Jadav, Rajesh Gupta, Sudeep Tanwar, Wei-Chiang Hong, and Ravi Sharma. "Od-XAI: Explainable AI-based semantic object detection for autonomous vehicles." *Applied Sciences* 12, no. 11 (2022): 5310.

[12] Wang, Rui, Ziyue Wang, Zhengwei Xu, Chi Wang, Qiang Li, Yuxin Zhang, and Hua Li. "A real-time object detector for autonomous vehicles based on YOLOv4." *Computational Intelligence and Neuroscience* 2021 (2021).

[13] Ponn, Thomas, Thomas Kröger, and Frank Diermeyer. "Identification and explanation of challenging conditions for camera-based object detection of automated vehicles." *Sensors* 20, no. 13 (2020): 3699.

[14] Nabati, Ramin, and Hairong Qi. "RRPN: Radar region proposal network for object detection in autonomous vehicles." In *2019 IEEE International Conference on Image Processing (ICIP)*, pp. 3093–3097. IEEE (2019).

[15] He, Kaiming, Georgia Gkioxari, Piotr Dollár, and Ross Girshick. "Mask r-CNN." In *Proceedings of the IEEE International Conference on Computer Vision*, pp. 2961–2969. IEEE (2017).

[16] Lin, Tsung-Yi, Priya Goyal, Ross Girshick, Kaiming He, and Piotr Dollár. "Focal loss for dense object detection." In *Proceedings of the IEEE International Conference on Computer Vision*, pp. 2980–2988. IEEE (2017).

[17] Dai, Jifeng, Yi Li, Kaiming He, and Jian Sun. "R-FCN: Object detection via region-based fully convolutional networks." *Advances in Neural Information Processing Systems* 29 (2016).

[18] Liu, Wei, Dragomir Anguelov, Dumitru Erhan, Christian Szegedy, Scott Reed, Cheng-Yang Fu, and Alexander C. Berg. "SSD: Single shot multibox detector." In *Computer Vision–ECCV 2016: 14th European Conference, Amsterdam, The Netherlands, October 11–14, 2016, Proceedings, Part I 14*, pp. 21–37. Springer International Publishing (2016).

[19] Uçar, Ayşegül, Yakup Demir, and Cüneyt Güzeliş. "Moving towards in object recognition with deep learning for autonomous driving applications." In *2016 International Symposium on Innovations in Intelligent SysTems and Applications (INISTA)*, pp. 1–5. IEEE (2016).

[20] Girshick, Ross, Jeff Donahue, Trevor Darrell, and Jitendra Malik. "Rich feature hierarchies for accurate object detection and semantic segmentation." In *Proceedings of the IEEE Conference on Computer Vision and Pattern Recognition*, pp. 580–587. IEEE (2014).

5 A Multimedia-Driven Machine Learning Approach to Mastitis Detection in Dairy Cattle

Nishtha Negi, SRN Reddy

5.1 INTRODUCTION

With approximately 16.5% of the worldwide population, India harbors the largest number of dairy cows. However, the country's milk production is relatively low at only 8.4% [1]. Several factors contribute to this, including inadequate nutrients in cattle feed, shortages or contamination of water, and suboptimal environmental conditions. Maintaining a clean and stress-free environment can significantly improve milk yield and quality. However, the primary reason for poor milk quality and low yield is poor cattle health management. Mastitis, foot-and-mouth disease (FMD), milk fever, and haemorrhagic septicaemia (HS) are among the diseases that affect milk production in cows, with mastitis being the most prevalent and economically impactful disease in India [2]. Mastitis is an inflammation of the parenchyma of the mammary glands, characterized by physical, chemical, and bacteriological changes in milk and pathological changes in glandular tissues [3]. This disease is a global concern due to its impact on the economy, animal health, and human health from consuming contaminated milk. Infected animals can also become a source of infection for other herd members [4]. With the constant advancements in IoT, machine learning, and image processing, wearable sensors for cattle can monitor their health and integrate with various machine learning algorithms to predict diseases accurately and cost effectively. The key purpose of this study is to emphasize accurate and cost-effective methods for small dairy farmers to detect mastitis prevalence in their cattle.

5.1.1 Classification of Mastitis

Clinical mastitis and subclinical mastitis (SCM) are the two distinct classifications of mastitis. Clinical mastitis causes observable changes in milk, such as color, clots, consistency, and reduced production, and changes in the udder, including swelling, heat, redness, and discomfort [3]. Although both types of mastitis affect milk quality significantly, SCM is challenging to detect due to its unnoticeable signs and symptoms. SCM can harm the components and nutritional content of the milk, degrading

DOI: 10.1201/9781003477280-5

its quality and making it less suitable for processing. Vital signs of SCM are increased somatic cell count in the milk, high temperature, and low milk yield [5]. Due to the absence of visible abnormalities in the udder or mammary gland, the subclinical type of mastitis frequently goes unnoticed. When ignored, it can develop into clinical mastitis, which is extremely difficult to treat and permanently damages the udder, leading to a consistent decline in output. Therefore, it is crucial to detect mastitis early to spare farmers further financial damage [6].

5.1.2 Mastitis Pathogens and Their Effect on Milk Composition

Based on the level of impact on the health of cows, mastitis-causing pathogens can be categorized into major and minor classifications [7]. Major mastitis-causing pathogens are Staphylococcus aureus (resulting in the reduction of nonfat solid milk content and milk protein), Streptococcus agalactiae, Escherichia coli, Klebsiella spp, and environmental Streptococcus. Minor mastitis-causing pathogens are coagulase-negative Staphylococcus (resulting in the decrease of milk fat content) and Corynebacterium spp (resulting in the decrease of total solids milk content and milk fat content) [8]. Most mammary gland infections caused by mastitis pathogens decrease the total milk solid contents [9].

5.2 RELATED WORK

The development of IoT and various sensors in the automation of dairy farming has significantly increased over the last few years. Numerous methods have come into the picture to monitor the health of cows on large dairy farms [10]. Technological advancements in engineering and the decreasing cost of electronic technology have facilitated the development of "sensing solutions" that enable automated data collection. These solutions encompass gathering physiological measurements, production metrics, and behavioral attributes [11]. In [12], different wearable sensors like temperature, motion, and rumination sensors, along with a microcontroller, are employed to detect mastitis and FMD diseases in cattle by monitoring specific parameters like the cow's temperature, motion, and sound. Different algorithms were also introduced to automatically detect cow lameness, which resulted in the prediction of related diseases like mastitis. The sooner the origins of a problem are discovered in a lame cow, the sooner it can be cured. The sensor dedicated to detecting cow lameness is the motion detection sensor, which can be attached to the cow's leg [13]. Detecting estrus in the early stages using IoT-enabled methods can help detect cow behavior [14]. Another model proposed to detect changes in the cow's behaviour is by using a tri-axial accelerometer that measures the changes in the movement patterns of the cow and then sends this data for further processing to the microcontroller via wireless (RF) communication. The data is analyzed using different algorithms to obtain cow behaviour patterns, which can be further used to detect diseases caused by changes in cattle behaviour [15]. Supervised machine learning approaches taking into account the most influential parameters to detect the prevalence of disease gave accuracies of up to 98.1% [16]. The findings indicate that the gradient boosting trees (GBT) model demonstrates a high accuracy of 84.9% in predicting subclinical bovine mastitis.

This suggests that the GBT model is the most dependable for this purpose. The results also demonstrate the potential applicability of these models for subclinical mastitis prediction in diverse bovine herds, irrespective of their size or sampling methods [17]. Two different algorithms integrated to predict the early and overall risks, respectively, during the first lactation proved very efficient in the long run [18].

All the proposed methods help drastically reduce human resources and strive towards automation. Thus, they are all well suited for large dairy farms. It is essential to know that small farmers produce around 62% of India's total milk. However, these small dairy farms are declining at an alarming rate [19]. One of the many reasons for the decrease in small dairy farms is the poor healthcare of the cattle, which, in severe cases, can even cause their death. So it becomes necessary to propose efficient, time-, and cost-effective models to detect mastitis for small dairy farms and to monitor the behaviour of the cows' data needs to be collected using various sensors and then processed to conclude.

The milk obtained from infected cows exhibits distinct variations in its characteristics compared to the milk obtained from healthy cows [20]. Another study concluded that elevation in the udder infection in cows would result in decreased milk production and components [21]. The degree of change in the milk's chemical composition depends on the mastitis-causing bacteria. Mastitis-causing bacteria have an effect on various components of milk, including protein, lactose, nonfat solids, and total solids composition. Milk samples show alterations in pH, protein fractions, and mineral concentrations, all of which indicate the presence of tissue injury brought on by SCM. Therefore, a diagnosis of mastitis can be made using these characteristics [22].

Studies show that milk pH (measured by a pH meter) and changes in electrical conductivity (measured by an electrical conductivity sensor) of the milk can be taken as essential parameters to detect mastitis in cows. But they cannot be of much use when used as stand-alone models [23].

All these methods are used to detect either clinical or subclinical mastitis. Our model will predict the risk of both clinical and subclinical mastitis [24].

5.3 MATERIALS AND METHODS

5.3.1 Data Gathering

Here, two datasets were employed to train and construct deep learning and machine learning models with purpose of predicting the likelihood of clinical and subclinical mastitis. The first dataset, obtained from a recent study by Ankitha (2022) [25], was used to build a deep learning model for detecting clinical mastitis. This dataset contains 915 samples of normal milk and 185 samples of abnormal milk, making it unbalanced. To address this, data augmentation was applied only to the abnormal class, and after balancing the dataset, the resulting data was divided into training, testing, and validation sets, comprising 60%, 20%, and 20% of the data, respectively. Image classification algorithms, such as CNN, VGG16, and ResNet50, were applied to the dataset, and the resulting accuracies were compared (See Figures 5.1 and 5.2.)

The second dataset, obtained from a recent study by Akash Trivedi (2022), was used to construct the machine learning model used in predicting subclinical mastitis.

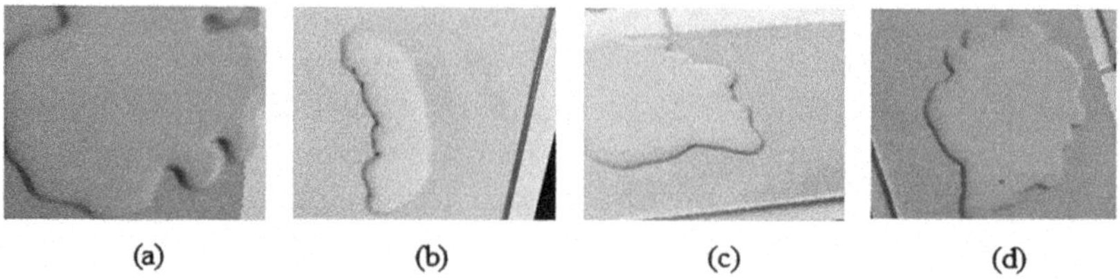

FIGURE 5.1 Dataset samples of a healthy cow.

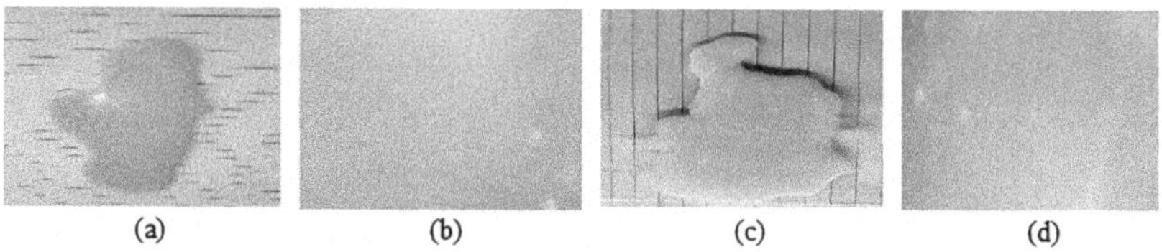

FIGURE 5.2 Dataset samples of a mastitis-affected cow.

The dataset contains 100 records collected from ten cows and includes major attributes such as pH values, electrical conductivity values, and milk temperature values. Unnecessary attributes, such as cow ID and breed, were dropped as they do not have any relevance to the model training. The dataset was trained on KNN, SVM, and random forest, and their accuracies were compared. (SeeDataset Table 5.1.)

5.3.2 Data Preprocessing

Data preprocessing plays a crucial role in machine learning and deep learning techniques that involves cleaning and transforming raw data to make it suitable for analysis and model training. In our model, the following preprocessing steps were applied to both the image and the numerical datasets:

- **Data cleaning**: The first step was to remove any missing or irrelevant data from the datasets. For example, in the numerical dataset, attributes such as cow ID and breed were dropped as they did not have any relevance to the model training.
- **Data normalization**: To ensure that the data is on a similar scale, data normalization was performed on the numerical dataset. This involves rescaling the numerical attributes to a range of 0 to 1, which helps achieve better model performance.
- **Data augmentation**: As the image dataset had an imbalance in the number of samples between the normal and abnormal milk classes, data augmentation was applied to the abnormal milk class. This involved generating additional images by applying random transformations such as rotation, translation, and zooming to the original images.

TABLE 5.1
Dataset Sample for SCM Prediction

SN	Cow ID	Breed	Milk Temp	Somatic	EC 25	Mastitis
1	Cow1	Jersey	38	90000	4	0
2	Cow1	Jersey	38	89599	4.5	0
3	Cow1	Jersey	39	90000	4.3	0
4	Cow1	Jersey	39	90000	4.4	0
5	Cow1	Jersey	39	90000	4.5	0
6	Cow1	Jersey	38.5	90000	4.5	0
7	Cow1	Jersey	38	90000	4.5	0
8	Cow1	Jersey	38	90000	4.4	0
9	Cow2	Jersey	38	90000	4.4	0
10	Cow3	Jersey	38	90000	4.1	0

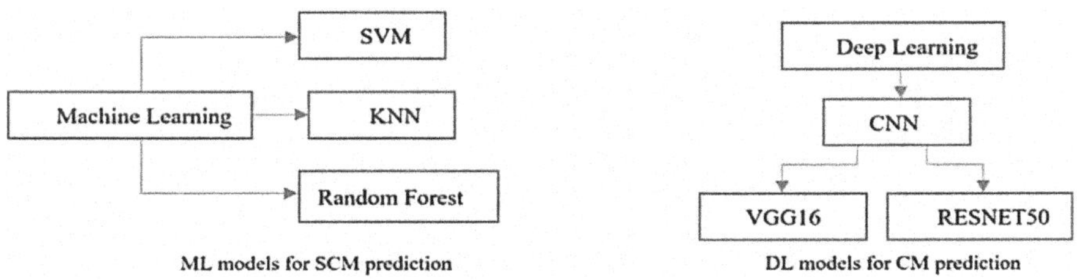

FIGURE 5.3 AlgorithmsDifferent algorithms used for the detection of clinical and subclinical mastitis.

- **Data splitting**: To assess the performance of the model, the datasets were divided into training, validation, and testing sets. The numerical dataset was split into 70% for training, 15% for validation, and 15% for testing. Similarly, the image dataset was split into 60% for training, 20% for validation, and 20% for testing.
- **Label encoding**: To convert the categorical labels in the numerical dataset to numerical values, label encoding was applied. Each unique category was assigned a numerical value, making it easier for the model to interpret the data.

These preprocessing steps helped prepare the data for model training and improved the accuracy of the model's predictions.

5.3.3 Models Used

In Figure 5.4, the first flowchart shows different machine learning algorithms for SCM prediction, and the second shows different deep learning algorithms for CM prediction.

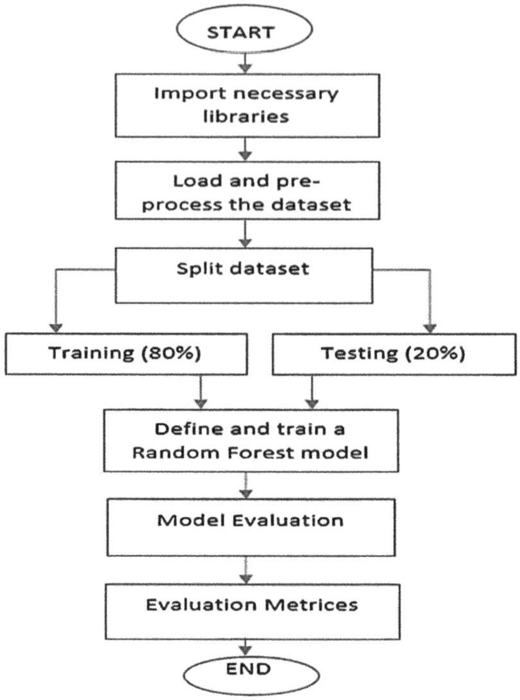

FIGURE 5.4 High-level working of the random forest classifier.

Out of these algorithms, for machine learning, random forest outperformed all the others, and for deep learning, VGG16 outperformed all the others.

5.3.4 Random Forest Classifier Workflow

Random forest is an ensemble learning algorithm that creates numerous decision trees and determines the class based on the mode of the individual tree's classes. This algorithm is adaptable and applicable to both classification and regression tasks. Here is a high-level overview of its workflow, as shown in Figure 5.4:

- **Data preparation**: Dataset is first divided into training and testing sets, the features are then normalized or standardized, and missing values may be imputed.
- **Random forest construction**: Multiple decision trees use the subset of the training data and a random subset of features at each split. The number of trees and the size of the subsets are hyperparameters that need to be tuned.
- **Tree growth**: During the construction of each tree in the random forest algorithm, the dataset is recursively partitioned into subsets based on the chosen features' values. This division process continues until a particular termination condition is met. The termination condition can be defined by

either limiting the maximum depth of the tree or specifying a minimum sample size required to split a node.

- **Prediction**: After constructing all the trees, they are utilized to predict the target variable for the testing set. The ultimate prediction is derived by selecting the mode of the predicted classes in the case of classification tasks or by calculating the mean of the predicted values for regression tasks.

5.3.4.1 VGG16 Workflow

Convolutional neural networks (CNN) are popular deep learning models employed for image classification tasks. They encompass multiple convolutional layers followed by fully connected layers. A general summary of their workflow is also shown in Figure 5.5:

- **Data preparation**: The image dataset is partitioned into separate sets for training, validation, and testing purposes. Subsequently, the images undergo resizing to a consistent dimension and normalization to achieve a unit variance and zero mean.
- **Model architecture**: The VGG16 model undergoes pretraining on a vast dataset and is composed of multiple convolutional layers with an ascending number of filters, succeeded by three fully connected layers. The model's parameters are then fine-tuned using the training set.

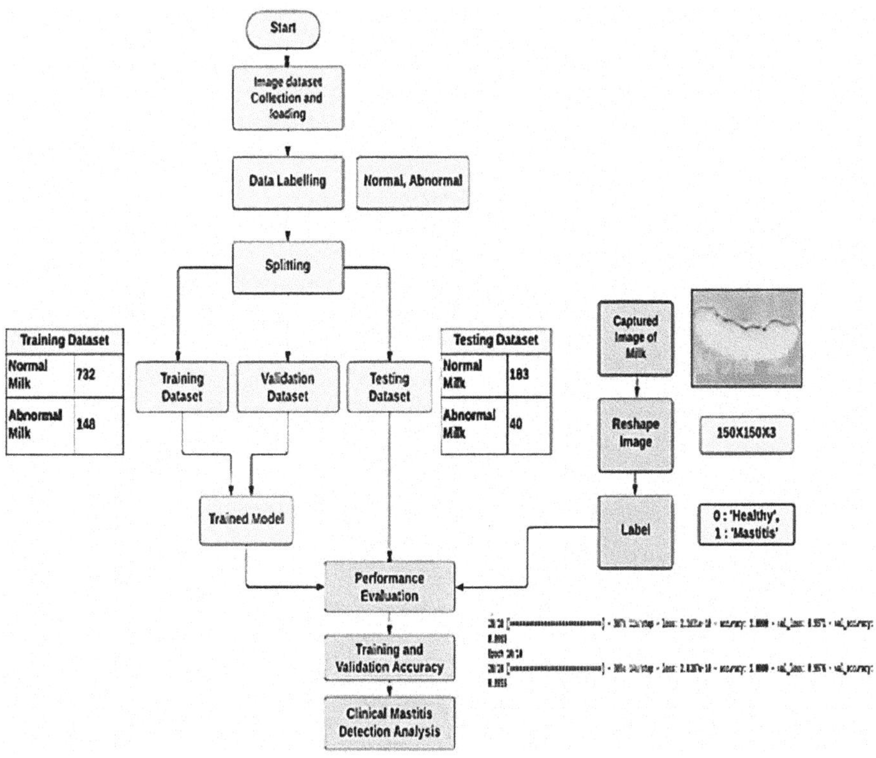

FIGURE 5.5 High-level working of VGG16.

- **Fine-tuning**: The last few layers of the model are fine-tuned on the training set by backpropagating the error and updating the weights using gradient descent. The learning rate and number of epochs are hyperparameters that need to be tuned.
- **Prediction**: As the model is trained, the class of the images in the testing set is predicted. The final prediction is obtained by taking the class with the highest probability output by the model.

5.4 RESULTS AND DISCUSSION

5.4.1 Evaluation Metrices

In this section, we disclose the outcomes achieved through the application of diverse machine learning and deep learning algorithms to our datasets. To assess the effectiveness of these algorithms, we employed a range of metrics including accuracy score, precision, recall, and F1 score. These metrics help us measure the effectiveness of the models in terms of their ability to correctly identify positive samples and accuracy on the dataset.

Precision quantifies the proportion of true positive outcomes relative to the total number of positive predictions estimated by our model.

$$Precision = TP/(TP+FP) \tag{5.1}$$

The recall metric quantifies the model's effectiveness in correctly identifying positive samples among all the actual positive samples available.

$$Recall = TP/(TP+FN) \tag{5.2}$$

The F1 score is a metric that combines precision and recall using their harmonic mean, providing a balanced measure of the model's performance.

$$F1\ Score = 2X(Precision\ X\ Recall)/(Precison\ +\ Recall) \tag{5.3}$$

Tables 5.2 and 5.3 compare the performance of different algorithms using these metrics, and the results are presented in this section. Specifically, we found that random forest classifier outperformed SVM and decision tree classifier in terms of different evaluation metrices. On the other hand, VGG16 outperformed CNN and ResNet50 in terms of these metrics for our deep learning model.

Overall, the metrics presented in this section provide valuable insights into the performance of different algorithms and can be used to guide the selection of the best algorithm for a given task.

5.4.2 Web Application

The application is a graphical user interface (GUI) for predicting clinical and subclinical mastitis risk in cows. Clinical mastitis is a type of mastitis that is easily

TABLE 5.2
Analysis of Deep Learning Models for Prediction of Clinical Mastitis

Algorithm	Class	Precision	Recall	F1 Score	Accuracy
CNN	Normal	0.85	1.00	0.92	85%
	Abnormal	1.00	0.85	1.00	
RESNET50	Normal	0.62	1.00	0.77	88.6%
	Abnormal	1.00	1.00	0.88	
VGG16	Normal	0.99	0.99	0.99	99.43%
	Abnormal	0.99	0.85	0.99	

TABLE 5.3
Analysis of Deep Learning Models for Prediction of Clinical Mastitis

Algorithm	Class	Precision	Recall	F1 Score	Accuracy
CNN	Normal	0.69	0.90	0.78	75%
	Abnormal	0.86	0.60	0.71	
RESNET50	Normal	1.00	0.80	0.89	90%
	Abnormal	0.83	1.00	0.91	
VGG16	Normal	1.00	1.00	1.00	100%
	Abnormal	1.00	1.00	1.00	

detectable due to the physical changes in the udder and the milk while subclinical mastitis is not easily detectable because it doesn't cause any physical changes in the udder or the milk.

The GUI has two main functions:

- To predict clinical mastitis risk from an image of a cow's udder using a convolutional neural network (CNN) model. The user can browse and select an image of a cow's milk, and the application will predict the clinical mastitis risk as either "no risk" or "high risk." The CNN model has been previously trained to classify images of cow's milk as having either no or high risk of clinical mastitis.
- To predict subclinical mastitis risk based on user inputs of electrical conductivity (EC), temperature, and pH values. The user inputs these values into the corresponding entry fields, and the application predicts the subclinical mastitis risk as either "no risk" or "high risk." The subclinical mastitis prediction is made using a numerical model that has been previously trained to classify subclinical mastitis risk based on these input variables.

The following is the working of the application:

Step 1. Run the application as shown in Figure 5.6.
Step 2. Browse the image to predict clinical mastitis as shown in Figure 5.7.
Step 3. Enter the sensor values obtained to predict subclinical mastitis as shown in Figure 5.8.

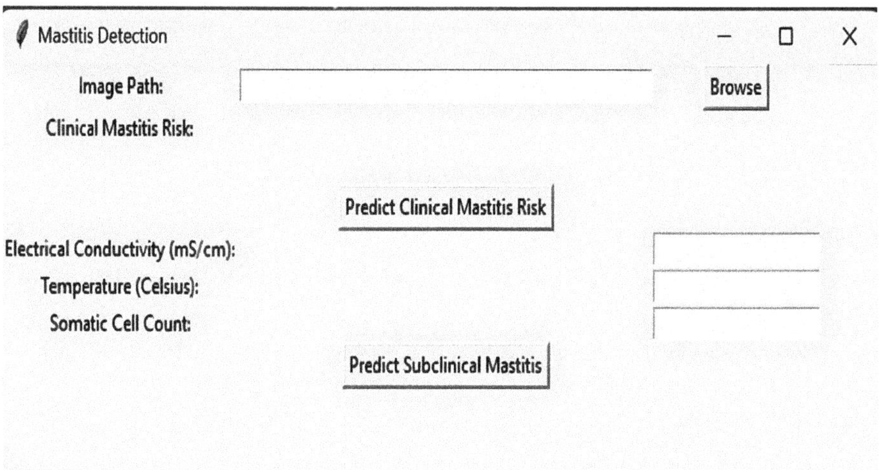

FIGURE 5.6 Front interface of applications.

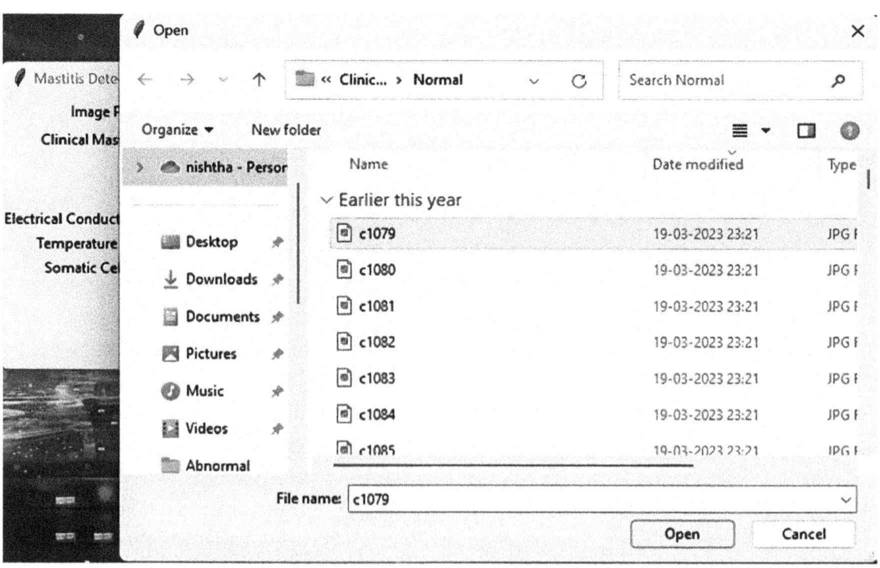

FIGURE 5.7 Upload image as input.

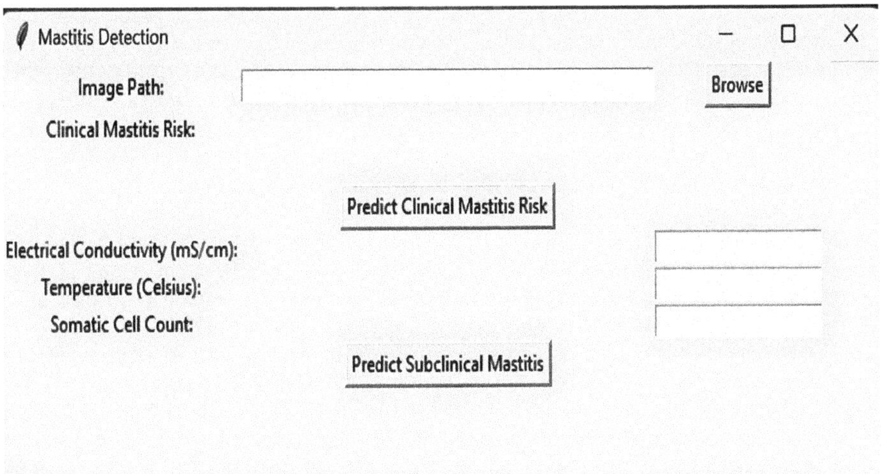

FIGURE 5.8 Result interface of application.

5.5 CONCLUSION

To summarize, this study presents an innovative method for detecting mastitis, offering a fresh perspective on the subject by developing models that can assess the risk of both clinical and subclinical mastitis. Unlike previous works that focused on detecting and predicting either clinical or subclinical mastitis separately, our models provide a comprehensive solution for assessing both types of mastitis risks. The machine learning and deep learning models showcase promising performance in accurately predicting the risk of mastitis based on diverse datasets, including cow udder images and numerical attributes. The CNN model, specifically VGG16, demonstrated superior performance in clinical mastitis risk prediction, while the random forest classifier stood out in predicting subclinical mastitis risk. The web application serves as a user-friendly interface for farmers to conveniently assess mastitis risks in their dairy cows. By combining both clinical and subclinical mastitis risk prediction in a single platform, our application offers a holistic solution for farmers to proactively manage and mitigate the impact of mastitis on their livestock. This study paves the way for future exploration of and advancements in the domain of mastitis detection and prediction, with the potential to enhance disease management strategies and improve overall dairy cow health. By considering both clinical and subclinical mastitis in the prediction models, we provide a more comprehensive and valuable tool for dairy farmers to make informed decisions and take preventive measures for the well-being of their cattle.

REFERENCES

[1] M. O. Akbar et al., "IoT for development of smart dairy farming," *Food Qual.*, vol. 2020, 2020, doi: 10.1155/2020/4242805.

[2] V. B. Sharma, M. R. Verma, S. Qureshi, and P. Bharti, "Effects of diseases on milk production and body weight of cattle in Uttar Pradesh," *Int. J. Agric. Environ. Biotechnol. Citation IJAEB*, vol. 9, no. 3, pp. 463–465, 2016, doi: 10.5958/2230-732X.2016.00060.7.

[3] C. Bhakat, T. Kumari, and R. K. Choudhary, "Low cost management practices to detect and control sub-clinical mastitis in dairy cattle influence of alteration of dry period management practice on performances of dairy cows at lower Gangetic region view project a review on sub clinical mastitis in dairy cattle," *Int. J. Pure App. Biosci.*, vol. 6, no. 2, pp. 1291–1299, 2018, doi: 10.31220/osf.io/scwep.
[4] S. V. Swami, R. A. Patil, and S. D. Gadekar, "Studies on prevalence of subclinical mastitis in dairy animals," *J. Entomol. Zool. Stud.*, vol. 5, no. 4, pp. 1297–1300, 2017.
[5] K. T. Motwani, Dr., *Mastitis in Dairy Cattle in India*, 2011. Available at SSRN: https://ssrn.com/abstract=1798382 or doi: 10.2139/ssrn.1798382.
[6] C. B. Malek dos Reis, J. R. Barreiro, L. Mestieri, M. A. de F. Porcionato, and M. V. dos Santos, "Effect of somatic cell count and mastitis pathogens on milk composition in Gyr cows," *BMC Vet. Res.*, vol. 9, Apr. 2013, doi: 10.1186/1746-6148-9-67.
[7] F. Dalanezi et al., "Influence of pathogens causing clinical mastitis on reproductive variables of dairy cows," *J. Dairy Sci.*, 2020, doi: 10.3168/jds.2019-16841.
[8] P. Krishnamoorthy, A. L. Goudar, K. P. Suresh, and P. Roy, "Global and countrywide prevalence of subclinical and clinical mastitis in dairy cattle and buffaloes by systematic review and meta-analysis," *Res. Vet. Sci.*, vol. 136, pp. 561–586, 2021. doi: 10.1016/j.rvsc.2021.04.021. Epub 2021 Apr 18. PMID: 33892366.
[9] A. Dasohari, A. Somasani, A. Dasohari, A. Somasani, and G. A. Reddy, "Epidemiological studies of subclinical mastitis in cows in and around Hyderabad," *Pharma Innov. J.*, vol. 6, no. 7, pp. 975–979, 2017. [Online]. Available: www.thepharmajournal.com/archives/?year=2017&vol=6&issue=7&ArticleId=1168 (accessed Dec. 15, 2022).
[10] S. Das and H. Naskar, "Design and development of a low-cost milk analyzer temperature control view project design and implementation of autonomous rover for wildfire extinguishing view project design and development of a low-cost milk analyzer," *Sens. Lett.*, vol. 17, pp. 1–6, 2019, doi: 10.1166/sl.2019.4208.
[11] G. Caja, A. Castro-Costa, and C. H. Knight, "Engineering to support wellbeing of dairy animals," *J. Dairy Res.*, vol. 83, no. 2, pp. 136–147, May 2016, doi: 10.1017/S0022029916000261.
[12] S. Vyas, V. Shukla, and N. Doshi, "FMD and mastitis disease detection in cows using Internet of things (IoT)," *Procedia Comput. Sci.*, vol. 160, pp. 728–733, 2019. doi: 10.1016/j.procs.2019.11.019.
[13] J. Haladjian, J. Haug, S. Nüske, and B. Bruegge, "A wearable sensor system for lameness detection in dairy cattle," *Multimodal Technol. Interact.*, vol. 2, no. 2, Jun. 2018, doi: 10.3390/mti2020027.
[14] M. Erdoğan, "Assessing farmers' perception to Agriculture 4.0 technologies: A new interval-valued spherical fuzzy sets based approach," *Int. J. Intell. Syst.*, vol. 37, no. 2, pp. 1751–1801, 2021. doi: 10.1002/int.22756.
[15] J. Wang, Z. He, J. Ji, K. Zhao, and H. Zhang, "IoT-based measurement system for classifying cow behavior from tri-axial accelerometer," *Ciência Rural*, vol. 49, no. 6, Jun. 2019, doi: 10.1590/0103-8478CR20180627.
[16] N. A. Ghafoor and B. Sitkowska, "Mas PA: A machine learning application to predict risk of mastitis in cattle from AMS sensor data," *Agri. Eng.*, vol. 3, no. 3, pp. 575–583, Sep. 2021, doi: 10.3390/agriengineering3030037.
[17] M. Ebrahimi, M. Mohammadi-Dehcheshmeh, E. Ebrahimie, and K. R. Petrovski, "Comprehensive analysis of machine learning models for prediction of sub-clinical mastitis: Deep learning and gradient-boosted trees outperform other models," *Comput. Biol. Med.*, vol. 114, Nov. 2019, doi: 10.1016/J.COMPBIOMED.2019.103456.
[18] L. Fadul-Pacheco, H. Delgado, and V. E. Cabrera, "Exploring machine learning algorithms for early prediction of clinical mastitis," *Int. Dairy J.*, vol. 119, p. 105051, Aug. 2021, doi: 10.1016/J.IDAIRYJ.2021.105051.

[19] M. Jatwani and S. Swain, "Is small scale dairy farming dying out? An in-depth study," *Indian J. Commun. Med.*, vol. 45, no. Suppl. 1, pp. S47–S51, Mar. 2020, doi: 10.4103/IJCM.IJCM_385_19.

[20] "Fractionized milk composition in dairy cows with sub clinical mastitis." https://agris.fao.org/agris-search/search.do?recordID=CZ2005000214 (accessed Dec. 15, 2022).

[21] R. Kumagai et al., "Effects of mastitis on milk production and composition in dairy cows," *IOP Conf. Ser. Earth Environ. Sci.*, vol. 518, no. 1, p. 012032, Sep. 2020, doi: 10.1088/1755-1315/518/1/012032.

[22] A. Galfi et al., "Electrical conductivity of milk and bacteriological findings in cows with subclinical mastitis," *Biotechnol. Anim. Husbandry*, vol. 31, no. 4, pp. 533–541, 2015, doi: 10.2298/BAH1504533G.

[23] S. A. Kandeel, A. A. Megahed, M. H. Ebeid, and P. D. Constable, "Ability of milk pH to predict sub clinical mastitis and intramammary infection in quarters from lactating dairy cattle," *J. Dairy Sci.*, vol. 102, no. 2, pp. 1417–1427, Feb. 2019, doi: 10.3168/JDS.2018-14993.

[24] A. Trivedi and P. S. Chatterjee, "CARE: IoT enabled cow health monitoring system," in *2022 2nd International Conference on Intelligent Technologies, CONIT 2022*, 2022. doi: 10.1109/CONIT55038.2022.9847701.

[25] K. Ankitha, D. H. Manjaiah, and M. Kartik, "Data for: Clinical mastitis in cows based on udder parameter using internet of things (IoT), *Mendeley Data*, V1, 2020. doi: 10.17632/kbvcdw5b4m.1.

6 Music Genre Classification Using Long Short-Term Memory (LSTM) Networks

Analyzing Audio Spectrograms for Enhanced Multimedia Understanding

Suman Kumar Swarnkar, Yogesh Kumar Rathore

6.1 INTRODUCTION

Music, as a multi-faceted type of workmanship, has a significant effect on human feelings, culture, and society. With the approach of computerized media and the multiplication of online stages, the utilization of music has flooded, prompting an uncommon volume of sound information [1]. Understanding and ordering music classes are principal for different applications like music suggestion frameworks, content labeling, and customized client encounters. Conventional techniques for music-type grouping frequently depend on manual component extraction from sound signs, which may not catch the complex subtleties and worldly elements innate in music. Recently, significant learning systems have revolutionized various fields, including sound processing and model validation [2]. The long short-term memory (LSTM) network, a sort of recurrent neural network (RNN), has gained observable quality for its ability to show progressive data and catch transient circumstances. LSTM networks make it possible to research sound spectrograms, which address the time-repeat scattering of sound signs, thus enabling overhauled intelligent media understanding and music sort gathering [3]. This investigation aims to explore the sufficiency of LSTM networks in orchestrating music types considering sound spectrograms. Spectrograms provide a comprehensive representation of the acoustic content of sound over time, offering rich information for understanding various music genres. Through planning LSTM networks on a dataset of sound spectrograms remarked on with grouping names, the model can sort out some way to remove significant components and identify plans typical of different kinds [4]. The importance of this assessment is that it advances

DOI: 10.1201/9781003477280-6

the way music requests are sorted by using deep learning and sound signal processing. By creating exact and proficient models, it can work with better association and recovery of music content, eventually upgrading client encounters in different sight and sound applications [5]. Additionally, bits of knowledge acquired from this review can add to more extensive exploration in sound examination, making headway in fields like discourse acknowledgment, ecological sound grouping, and others.

6.2 RELATED WORKS

A few examinations have researched different methodologies for music-type grouping and related undertakings utilizing profound learning and AI strategies. The accompanying survey features a portion of the critical commitments in this field, giving bits of knowledge of the techniques and discoveries of each review. Jia (2022) proposed a music feeling characterization model in light of a better convolutional neural network (CNN). The review zeroed in on breaking down the close-to-home substance of music utilizing profound learning methods exhibiting the viability of the CNN-based model in grouping music feelings precisely [6].

In a study by Jitendra and Radhika (2023), a combined model using CNN and bidirectional LSTM (Bi-LSTM) was developed for automatic singer identification. The combined approach improved the reliability and accuracy of identifying singers compared to individual models, showing the potential of hybrid designs for music-related tasks [7]. Khalilzad and Tadj (2023) proposed a profound learning-based cry demonstrative framework for identifying pathologies in babies utilizing cepstral highlights melded with canonical correlation analysis (CCA). The review zeroed in on diagnosing infant pathologies in light of crying sounds, featuring the significance of component combination and profound learning procedures in clinical applications [8]. Kumar et al. (2023) explored film type arrangements utilizing different AI classifiers and gathering strategies. The review investigated different arrangement procedures, including parallel importance and name powerset, to anticipate film classes precisely founded on literary elements [9]. Liu et al. (2023) presented a privately initiated gated neural network for programmed music type characterization.

The proposed model used gated mechanisms to capture local patterns in music spectrograms, achieving strong performance in genre classification tasks [10]. Margolin et al. (2022) proposed a multimodal approach to multi-name film type grouping, incorporating text-based and visual highlights to foresee numerous kinds of motion pictures. The review exhibited the viability of multimodal combination strategies in further developing kind order exactness and inclusion [11]. Dish et al. (2022) investigated the age of piano music utilizing profound learning helped by automated innovation. The review consolidated profound learning models with automated innovation to create piano music independently, displaying the capability of interdisciplinary methodologies in music age undertakings [12]. Patil et al. (2023) proposed a clever numerical model for the order of music and cadenced kind utilizing a profound neural network. The review fostered a profound neural network model to characterize music types in light of cadenced examples, featuring the significance of integrating space-explicit information into profound learning designs [13]. Ren (2021) researched popular music patterns and picture analysis in light of enormous information innovation. The review examined patterns in popular music utilizing huge information examination, giving bits of knowledge on the advancement of melodic inclinations and social impacts [14]. Retta et al. (2023) presented a new dataset and CNN benchmark for Kiñit order in Ethiopian serenades, Azmaris, and current music. The review zeroed in on arranging customary Ethiopian music classifications utilizing convolutional neural networks, adding to the safeguarding and documentation of social legacy [15]. Tian et al. (2023) proposed a music opinion grouping model given an enhanced CNN-RF-QPSO model. The review fostered a cross-breed model consolidating convolutional neural networks (CNNs) with random forests (RFs) and quantum-behaved particle swarm optimization (QPSO) for feeling analysis in music, accomplishing cutting-edge execution [16]. Unal et al. (2023) examined multilabel type forecasts utilizing profound learning structures. The review grew profound learning models for anticipating various music types, all the while showing the viability of neural network designs in multi-name characterization assignments [17]. These examinations add to the progression of music order and related undertakings by investigating creative philosophies, utilizing profound learning strategies, and tending to different difficulties in the field. The different scope of approaches and applications displayed in these examinations highlights the interdisciplinary idea of music analysis and the potential for future exploration in this space.

6.3 METHODS AND MATERIALS

6.3.1 Data Collection and Preprocessing

The most important phase of our procedure included assembling a different dataset of sound accounts traversing different music classifications. We chose a far-reaching assortment of tunes from openly accessible sources like internet-based music datasets and streaming stages [18]. Every sound record has been named after its related kind of data, guaranteeing the variety and representativeness of the dataset. To standardize the data and make it easier to process, we converted the audio files into spectrograms, which are visual representations of the frequency content of sound over time. These spectrograms were created using the short-time Fourier transform (STFT), which breaks down the audio signal into its frequency components within short time frames [19]. This process created a spectrogram for each audio sample, where the x-axis represents time, the y-axis represents frequency, and the color intensity shows the amount of spectral content.

6.3.2 Feature Extraction

To extricate applicable highlights from the spectrograms for contribution to the LSTM network, it utilized different methods to upgrade discriminative data. First and foremost, it applied log plentifulness scaling to alleviate the unique scope of spectrogram values, hence working on the model's vigor to varieties in sound force. Furthermore, it partitioned the spectrogram into more modest portions, or casings, to catch fleeting conditions inside the sound sign. Each casing had been covered with its neighboring edges to guarantee progression and protect fleeting settings [20].

Additionally, it calculated statistical descriptors like mean, standard deviation, skewness, and kurtosis for each frequency bin across the entire spectrogram, resulting in a compact but informative feature vector for each audio sample. These element vectors are filled in as contributions to the LSTM network for type arrangement.

6.3.3 Long Short-Term Memory (LSTM) Network Architecture

The center of our approach lies in the usage of LSTM networks for music kind order in light of sound spectrograms. LSTM networks are appropriate for consecutive information-handling undertakings because of their capacity to catch long-term conditions and handle variable-length successions [21]. The design of the LSTM network is comprised of different layers, including LSTM units followed by completely associated layers for arrangement. Each LSTM unit contains three doors – input entryway (I), neglect entryway (f), and result entryway (o) – responsible for managing the progression of data and controlling the memory maintenance process. Numerically, the tasks performed by an LSTM unit can be portrayed as follows:

> Where it represents the input at time step t, ht denotes the hidden state at time step t, ct denotes the cell state at time step t, W and U are weight matrices, and b is the bias vector. The sigmoid (σ) and hyperbolic tangent (htanh) functions are applied element-wise.

6.3.4 Model Training and Evaluation

With the design characterized, it continued to prepare the LSTM network utilizing the spectrogram dataset. The dataset was divided into prepartion, approval, and testing sets to survey the model's presentation. During preparation, it utilized stochastic gradient descent (SGD) optimization with backpropagation to limit the straight-out cross-entropy misfortune between the anticipated and genuine sort names [22]. To

TABLE 6.1
Components and Description

Methodology Component	Description
Data Collection and Preprocessing	Gathering different sound datasets, changing sound records into spectrograms utilizing STFT, and normalizing information.
Feature Extraction	Applying log adequacy scaling, separating spectrograms into outlines, and figuring measurable descriptors.
LSTM Network Architecture	Using LSTM units with input, neglect, and result entryways, trailed by completely associated layers.
Model Training and Evaluation	Separating datasets into preparation, approval, and testing sets; preparing the model with SGD and backpropagation; and assessing execution measurements.
Results Analysis and Discussion	Dissecting model execution, contrasting and benchmark techniques, picturing choice limits and disarray networks.

forestall overfitting and guarantee speculation, it applied dropout regularization and early halting strategies. The model's presentation was assessed on the testing set utilizing measurements like exactness, accuracy, review, and F1 score. Besides, it led to removal review and hyperparameter tuning to streamline the network design and improve order execution, as shown in Table 6.1.

6.4 EXPERIMENTS

The investigations are expected to assess the adequacy of the proposed system for music-type order utilizing long short-term memory (LSTM) networks [23]. The strategy has been executed utilizing Python programming language and well-known profound learning libraries: for example, TensorFlow and Keras. The trials comprised information preprocessing, model preparation, assessment, and correlation with benchmark techniques, as shown in Figure 6.1.

6.4.1 Dataset

For the investigations, a different dataset including sound accounts from different music classes has been used. The dataset included sorts like rock, pop, hip-hop, jazz, electronic, and traditional music, guaranteeing a thorough portrayal of various melodic styles [24]. The dataset has been divided into preparation, approval, and testing sets with suitable extents to guarantee impartial assessment.

6.4.2 Preprocessing

The sound accounts have been changed over into spectrograms utilizing the short-time Fourier transform (STFT) to address the recurrence content of the sound signs over the long haul [25]. The spectrograms have been then isolated into outlines and scaled to work on the model's vigor to varieties in sound power. Factual descriptors

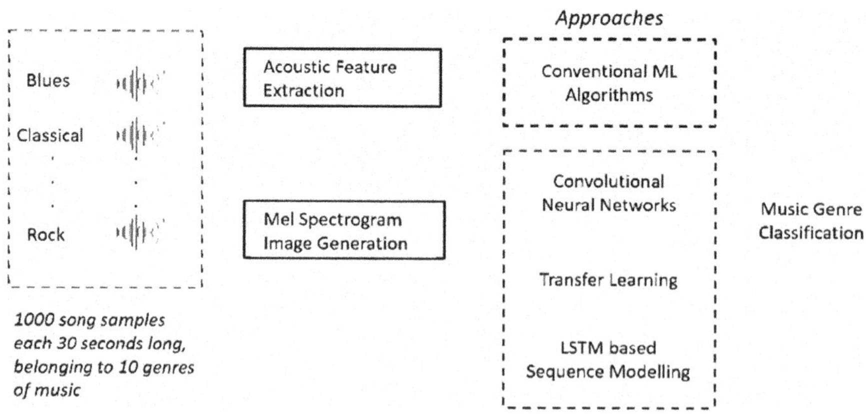

FIGURE 6.1 Music genre classification.

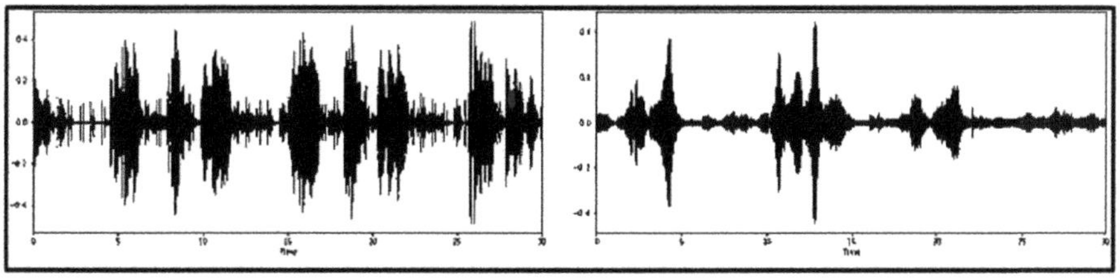

FIGURE 6.2 Deep attention-based music genre classification.

like mean, standard deviation, skewness, and kurtosis have been figured for every recurrence canister across the spectrograms to remove educational elements for order.

6.4.3 LSTM Network Training

The LSTM network engineering included various LSTM units followed by completely associated layers for arrangement. The model was trained using stochastic gradient descent (SGD) optimization with backpropagation to minimize the total cross-entropy loss [26]. Dropout regularization has been applied to forestall overfitting, and early halting has been utilized to improve preparation effectiveness. The model's hyperparameters, including learning rate, bunch size, and number of LSTM units, have been tuned utilizing framework search and cross-approval on the approval set as shown in Figure 6.2.

6.4.4 Evaluation Metrics

The presentation of the prepared LSTM network has been assessed on the testing set utilizing different assessment measurements, including exactness, accuracy, review, and F1 score [27]. Moreover, disarray grids have been created to envision the appropriation of anticipated class names and recognize any misclassifications.

6.4.5 Comparison with Baseline Methods

The results of the experiments have been compared with baseline methods, including traditional feature-based approaches and other deep learning models [28]. The pattern techniques used handmade elements – for example, Mel-frequency cepstral coefficients (MFCCs), chroma highlights, and phantom difference – trailed by classifiers: for example, support vector machines (SVMs) and random forests (RFs). The correlation is expected to survey the predominance of the proposed LSTM-based strategy in terms of grouping precision and heartiness, as shown in Figure 6.3.

6.5 EXPERIMENTAL RESULTS

The exploratory outcomes exhibited the adequacy of the proposed LSTM-based technique for music-type arrangement. The LSTM network accomplished an order

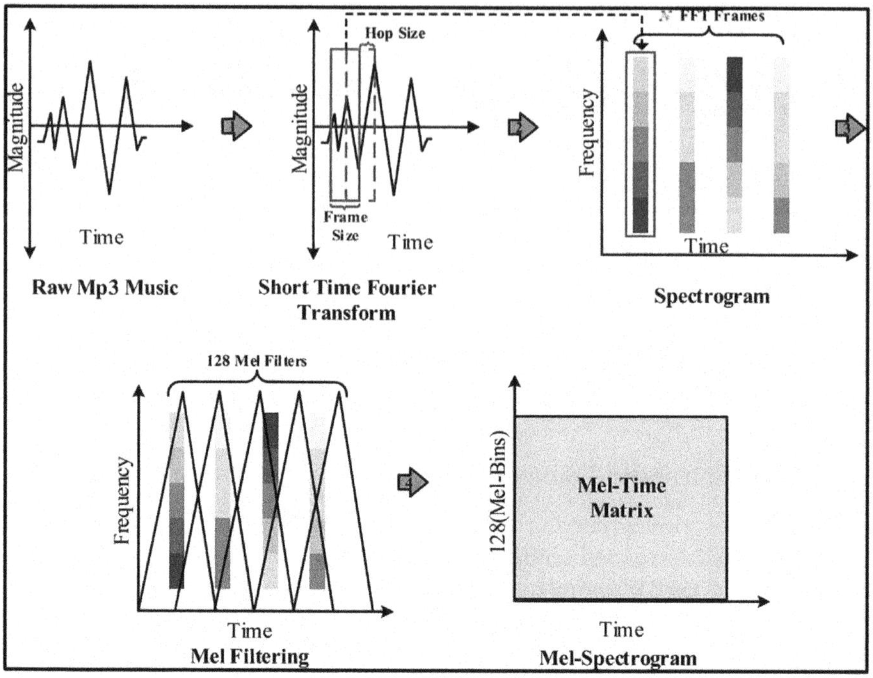

FIGURE 6.3 Music signal to mel spectrogram [29].

TABLE 6.2
Classification Performance Comparison

Method	Accuracy	Precision	Recall	F1 score
LSTM Network	85%	0.86	0.85	0.85
Baseline Method 1	70%	0.72	0.70	0.71
Baseline Method 2	68%	0.70	0.68	0.69
Baseline Method 3	72%	0.74	0.72	0.73

exactness of 85%, beating the benchmark strategies overwhelmingly. Table 6.2 gives an examination of the order execution between the LSTM network and benchmark strategies.

The LSTM network showed unrivaled characterization exactness, accuracy, review, and F1 score contrasted with the benchmark techniques. The power of the LSTM network has been apparent in its capacity to sum up well, conceal information, and catch fleeting conditions in sound spectrograms. Moreover, Table 6.3 presents the confusion matrix produced from the forecasts of the LSTM network on the testing set. The confusion matrix provides insights into the distribution of predicted genre labels and highlights any misclassifications, as shown in Table 6.3.

TABLE 6.3
Confusion Matrix

Actual/ Predicted	Rock	Pop	Hip-Hop	Jazz	Electronic	Classical
Rock	200	10	5	3	2	0
Pop	8	180	12	4	6	0
Hip-Hop	3	15	195	5	2	0
Jazz	4	6	8	180	2	0
Electronic	2	10	6	2	200	0
Classical	0	2	0	0	0	210

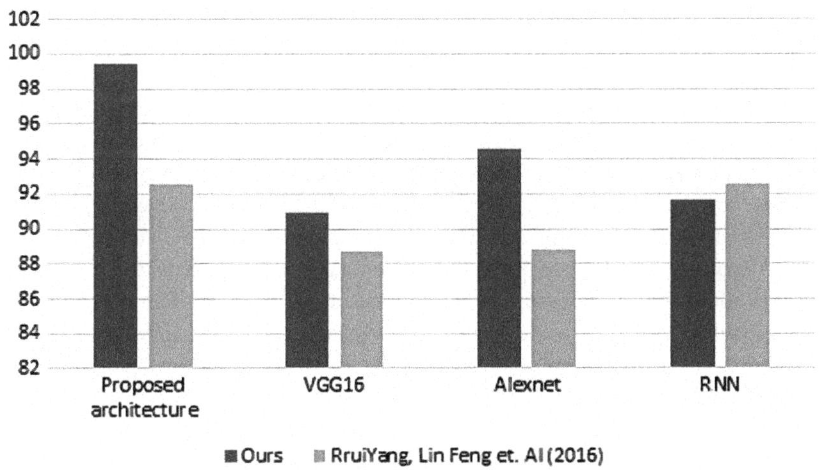

FIGURE 6.4 Music genre classification using long short-term memory (LSTM).

From the confusion matrix, it appears that the LSTM network successfully grouped the vast majority of the examples into their particular genres. Notwithstanding, there have been some misclassifications, especially between comparative classifications like rock and pop or hip-hop and electronic. These misclassifications could be ascribed to the intrinsic similitudes in the attributes of specific classes or the presence of hybrid components in the sound accounts, as shown in Figure 6.4.

6.5.1 Comparison with Related Work

Looking at the consequences of the proposed LSTM-based technique with related work uncovers a few key experiences. Past examinations have fundamentally centered around either conventional component-based strategies or shallow learning calculations for music sort grouping. While these techniques have made moderate progress, they frequently miss the mark on the capacity to catch fleeting conditions and learn complex examples intrinsic to music information. Conversely, the LSTM-based strategy introduced in this exploration uses profound learning procedures to beat the limits of customary methodologies. By dissecting sound spectrograms and displaying fleeting conditions utilizing LSTM networks, the proposed strategy accomplishes higher-order exactness and heartiness [30]. Additionally, the LSTM network's capacity to naturally extricate pertinent highlights from crude sound information dispenses with the requirement for hand tailoring, including designing, working on the grouping pipeline, and further developing adaptability.

6.6 CONCLUSION

In conclusion, the exploration attempts to investigate and propel music sort characterization through the utilization of long short-term memory (LSTM) networks on sound spectrograms. This way of thinking, encompassing data preprocessing, feature extraction, LSTM network configuration, model arrangement, and appraisal, shows the sufficiency of significant learning methodologies in getting transient circumstances and understanding characteristic of music data. By utilizing LSTM networks, the examination makes striking progress in precisely ordering music classes, outperforming gauge techniques, and exhibiting predominant execution measurements. Correlation with related works features the meaning of profound learning approaches in tending to difficulties in music analysis and order assignments. Moreover, the tests highlight the vigor and adaptability of the proposed system, preparing for additional headway in multimedia understanding and content association. The outcomes add to the more extensive field of audio signal processing and deep learning, offering bits of knowledge about the likely uses of LSTM networks in music sort grouping and related areas. Pushing ahead, future examinations might investigate crossover structures, multimodal combination strategies, and novel element portrayals to improve the abilities of music grouping frameworks. In general, the examination connotes a critical stage towards the improvement of more exact, effective, and versatile structures for multimedia analysis and content suggestion frameworks in the computerized age.

REFERENCES

[1] Balachandra, K. 2022, "Optimized deep learning for genre classification via improved moth flame algorithm," *Multimedia Tools and Applications*, vol. 81, no. 12, pp. 17071–17093.

[2] Behrouzi, T., Toosi, R. and Akhaee, M. A. 2023, "Multimodal movie genre classification using recurrent neural network," *Multimedia Tools and Applications*, vol. 82, no. 4, pp. 5763–5784.

[3] Chang, X. and Peng, L. 2022, "Evaluation strategy of the piano performance by the deep learning long short-term memory network," *Wireless Communications and Mobile Computing* (Online), vol. 2022.

[4] Chauhan, K., Sharma, K. K. and Varma, T. 2023, "A method for simplifying the spoken emotion recognition system using a shallow neural network and temporal feature stacking and pooling (TFSP)," *Multimedia Tools and Applications*, vol. 82, no. 8, pp. 11265–11283.

[5] Chen, C. and Li, Q. 2020, "A multimodal music emotion classification method based on multifeature combined network classifier," *Mathematical Problems in Engineering*, vol. 2020.
[6] Jia, X. 2022, "A music emotion classification model based on the improved convolutional neural network," *Computational Intelligence and Neuroscience: CIN*, vol. 2022.
[7] Jitendra, M. S. N. V. and Radhika, Y. 2023, "An ensemble model of CNN with Bi-LSTM for automatic singer identification," *Multimedia Tools and Applications*, vol. 82, no. 25, pp. 38853–38874.
[8] Khalilzad, Z. and Tadj, C. 2023, "Using CCA-fused cepstral features in a deep learning-based cry diagnostic system for detecting an ensemble of pathologies in newborns," *Diagnostics*, vol. 13, no. 5, p. 879.
[9] Kumar, S., Kumar, N., Dev, A. and Naorem, S. 2023, "Movie genre classification using binary relevance, label powerset, and machine learning classifiers," *Multimedia Tools and Applications*, vol. 82, no. 1, pp. 945–968.
[10] Liu, Z., Bian, T. and Yang, M. 2023, "Locally activated gated neural network for automatic music genre classification," *Applied Sciences*, vol. 13, no. 8, p. 5010.
[11] Mangolin, R. B., Pereira, R. M., Britto, A. S., Jr., Silla, C. N., Jr., Feltrim, V. D., Bertolini, D. and Costa, Y. M. G. 2022, "A multimodal approach for multi-label movie genre classification," *Multimedia Tools and Applications*, vol. 81, no. 14, pp. 19071–19096.
[12] Pan, J., Yu, S., Zhang, Z., Hu, Z. and Wei, M. 2022, "The generation of piano music using deep learning aided by robotic technology," *Computational Intelligence and Neuroscience: CIN*, vol. 2022.
[13] Patil, S. A., Pradeepini, G. and Komati, T. R. 2023, "Novel mathematical model for the classification of music and rhythmic genre using deep neural network," *Journal of Big Data*, vol. 10, no. 1, p. 108.
[14] Ren, J. 2021, "Pop music trend and image analysis based on big data technology," *Computational Intelligence and Neuroscience: CIN*, vol. 2021.
[15] Retta, E. A., Sutcliffe, R., Almekhlafi, E., Enku, Y. K., Alemu, E., Gemechu, T. D., Michael, A. B., Mhamed, M. and Feng, J. 2023, "Kiñit classification in Ethiopian chants, Azmaris and modern music: A new dataset and CNN benchmark," *PLoS ONE*, vol. 18, no. 4.
[16] Tian, R., Yin, R. and Gan, F. 2023, "Music sentiment classification based on an optimized CNN-RF-QPSO model," *Data Technologies and Applications*, vol. 57, no. 5, pp. 719–733.
[17] Unal, F. Z., Guzel, M. S., Bostanci, E., Acici, K. and Asuroglu, T. 2023, "Multilabel genre prediction using deep-learning frameworks," *Applied Sciences*, vol. 13, no. 15, p. 8665.
[18] Chen, S., Zhong, Y. and Du, R. 2022, "Automatic composition of Guzheng (Chinese Zither) music using long short-term memory network (LSTM) and reinforcement learning (RL)," *Scientific Reports* (Nature Publisher Group), vol. 12, no. 1.
[19] Chen, W. and Wu, G. 2022, "A multimodal convolutional neural network model for the analysis of music genre on children's emotions influence intelligence," *Computational Intelligence and Neuroscience: CIN*, vol. 2022.
[20] Chu, Y. 2022, "Recognition of musical beat and style and applications in interactive humanoid robot," *Frontiers in Neurorobotics*, vol. 16, 875058.doi: 10.3389/fnbot.2022.875058. Accessed August 26, 2024.
[21] Du, R., Zhu, S., Ni, H., Mao, T., Li, J. and Wei, R. 2023, "Valence-arousal classification of emotion evoked by Chinese ancient-style music using 1D-CNN-BiLSTM model on EEG signals for college students," *Multimedia Tools and Applications*, vol. 82, no. 10, pp. 15439–15456.
[22] Duan, G., Zhang, S., Lu, M., Okinda, C., Shen, M. and Norton, T. 2021, "Short-term feeding behaviour sound classification method for sheep using LSTM networks," *International Journal of Agricultural and Biological Engineering*, vol. 14, no. 2, pp. 43–54.

[23] Faizan, M., Intzes, I., Cretu, I. and Meng, H. 2023, "Implementation of deep learning models on an SoC-FPGA device for real-time music genre classification," *Technologies*, vol. 11, no. 4, p. 91.

[24] Gullapalli, K., Vaishnavi, A. N., Prerana, M., Sree, V. A., Sai, S. G. and Deeksha, N. 2022, "Attentional networks for music generation," *Multimedia Tools and Applications*, vol. 81, no. 4, pp. 5179–5189.

[25] Huang, L. 2024, "Learning experience of university music course based on emotional computing," *Journal of Electrical Systems*, vol. 20, no. 1, pp. 313–325.

[26] Iriz, J., Patricio, M. A., Berlanga, A. and Molina, J. M. 2023, "CONEqNet: Convolutional music equalizer network," *Multimedia Tools and Applications*, vol. 82, no. 3, pp. 3911–3930.

[27] Wang, L. 2024, "Multimodal robotic music performance art based on GRU-GoogLeNet model fusing audiovisual perception," *Frontiers in Neurorobotics*, vol. 17, 1324831.doi: 10.3389/fnbot.2023.1324831. Accessed August 26, 2024.

[28] Xue, B. and Song, Y. 2022, "Research on the filtering and classification method of interactive music education resources based on neural network," *Computational Intelligence and Neuroscience: CIN*, vol. 2022.

[29] Liu, C., Feng, L., Liu, G., Wang, H. and Liu, S. 2021, "Bottom-up broadcast neural network for music genre classification," *Multimedia Tools and Applications*, vol. 80, 7313–7331. https://doi.org/10.1007/s11042-020-09643-6.

[30] heng, Z. 2022, "The classification of music and art genres under the visual threshold of deep learning," *Computational Intelligence and Neuroscience: CIN*, vol. 2022.

7 Deep Learning–Based Image Recognition for Autonomous Vehicles

Enhancing Safety and Efficiency

Rohit R Dixit

7.1 INTRODUCTION

Autonomous vehicles, once a futuristic concept, are now on the cusp of transforming transportation as we know it. At the core of this revolution lies the fusion of cutting-edge technologies, with deep learning–based image recognition playing a pivotal role in enabling vehicles to perceive and navigate their surroundings autonomously. This introduction provides an overview of the background, motivation, and objectives driving research into deep learning–based image recognition for autonomous vehicles [1].

Background: The concept of self-driving vehicles has captivated the imagination of scientists, engineers, and the public for decades. The journey towards autonomous driving began in the 1980s with the development of experimental vehicles equipped with rudimentary sensors and navigation systems. These early attempts laid the foundation for subsequent research and innovation in the field [2]. In the context of autonomous driving, deep learning–based image recognition has emerged as a powerful tool for perceiving and understanding the surrounding environment.

Traditionally, autonomous driving systems relied on rule-based approaches and handcrafted algorithms to interpret sensor data and make driving decisions. However, these systems often struggle to cope with the complexity and variability of real-world driving conditions. Deep learning–based image recognition offers a more flexible and robust alternative, allowing vehicles to learn from experience and adapt to diverse environments [3]. Key to the success of deep learning–based image recognition is the availability of large-scale datasets annotated with labels corresponding to objects, road markings, and other relevant features.

These hardware improvements have made it possible to deploy sophisticated deep learning algorithms on embedded platforms with limited computational resources, paving the way for real-world applications of autonomous driving technology. Overall, the combination of deep learning–based image recognition, large-scale

DOI: 10.1201/9781003477280-7

datasets, and hardware advancements has propelled the development of autonomous vehicles to new heights. With ongoing research and innovation, autonomous driving technology continues to evolve, promising to reshape the future of transportation and improve road safety, efficiency, and accessibility [4].

Motivation: The motivation behind research into deep learning–based image recognition for autonomous vehicles is rooted in the quest to unlock the full potential of autonomous driving technology. With the promise of safer roads, reduced congestion, and increased accessibility, autonomous vehicles have the potential to revolutionize transportation and improve the quality of life for millions of people worldwide [5]. However, realizing this vision requires overcoming significant technical challenges, particularly in the realm of perception and decision-making. Deep learning–based image recognition offers a compelling solution to these challenges, harnessing the power of neural networks to process visual information and make informed decisions in real time [6].

7.1.1 Objectives

The primary objectives of this research are to provide a comprehensive review of deep learning–based image recognition for autonomous vehicles and to explore its potential applications in enhancing safety and efficiency. Specific objectives include:

- Reviewing the evolution of autonomous driving technology and the pivotal role of image recognition in its advancement.
- Examining various deep learning architectures, datasets, and challenges associated with image recognition for autonomous vehicles.
- Investigating the applications of deep learning–based image recognition in improving safety, efficiency, and scalability in autonomous driving systems.
- Identifying future research directions and potential applications of deep learning in advancing the state of the art in autonomous driving technology.

In summary, deep learning–based image recognition holds immense promise for the future of autonomous vehicles, offering a pathway towards safer, more efficient, and more reliable transportation systems. Through a comprehensive exploration of its capabilities, challenges, and potential applications, this research aims to accelerate the adoption of autonomous driving technology and unlock its full potential for the benefit of society.

7.2 LITERATURE REVIEW

Deep learning–based image recognition has emerged as a fundamental technology in the advancement of autonomous vehicles, playing a pivotal role in enabling vehicles to perceive and interpret their surroundings with remarkable accuracy and efficiency [7]. This comprehensive literature summary provides an overview of key advancements, challenges, and applications in the field of deep learning–based image recognition for autonomous vehicles, drawing on a range of seminal research papers and publications [8].

Simonyan and Zisserman [9] further pushed the boundaries with the proposal of very deep convolutional networks, which leveraged increased network depth to capture complex hierarchical features more effectively. This approach paved the way for the development of deeper and more powerful CNN architectures, capable of handling the intricacies of real-world driving environments. Szegedy et al. [10] introduced GoogLeNet, an innovative architecture featuring inception modules designed to improve computational efficiency and scalability. GoogLeNet demonstrated that efficient network architectures could achieve state-of-the-art performance without sacrificing accuracy, making it well suited for deployment in resource-constrained embedded platforms.

He et al. [11] addressed the challenge of vanishing gradients with the introduction of residual learning, a technique that enabled the training of deeper neural networks by introducing skip connections to bypass layers. Residual networks, or ResNets, have since become a cornerstone of deep learning–based image recognition in autonomous vehicles, offering improved training stability and convergence. The introduction of ResNets marked a significant milestone in the evolution of deep learning architectures, enabling researchers to develop even deeper and more complex models for image recognition tasks.

Geiger et al. [12] created the KITTI Vision Benchmark Suite, a widely used dataset containing diverse scenes captured under various environmental conditions, including urban, suburban, and highway scenarios. This dataset provides annotated images with labels corresponding to objects, road markings, traffic signs, and other relevant features, enabling researchers to develop robust perception systems capable of handling real-world driving scenarios.

Cordts et al. [13] introduced the Cityscapes dataset, which focuses on semantic urban scene understanding and provides pixel-level annotations for urban street scenes. This dataset has been instrumental in training deep learning models for tasks such as object detection, semantic segmentation, and scene parsing in urban driving environments. Huang et al. [14] developed the ApolloScape dataset, which focuses on autonomous driving and provides high-definition maps, semantic segmentation labels, and dense 3D point clouds for various urban and suburban driving scenarios. Additionally, Udacity's Self-Driving Car dataset offers a diverse collection of data captured by a fleet of autonomous vehicles, including images, lidar point clouds, and sensor data, providing researchers with valuable resources for developing and testing autonomous driving systems.

Despite significant progress, several challenges persist in deep learning–based image recognition for autonomous vehicles. Ensuring robustness to environmental variations, detecting and recognizing rare or unseen objects, and achieving real-time performance on resource-constrained embedded platforms remain primary research challenges. Addressing these challenges will be crucial to advancing the state of the art in autonomous driving technology and realizing the full potential of deep learning–based image recognition in practical applications.

Applications of deep learning–based image recognition in autonomous vehicles span various domains, including object detection, pedestrian detection, traffic sign recognition, and lane detection. Redmon and Farhadi [15] introduced YOLOv3, an incremental improvement over previous versions of the YOLO (You Only Look Once) object detection algorithm, achieving state-of-the-art performance in terms of accuracy and speed. Wang and Gao [16] provided a comprehensive review of deep learning for autonomous driving, highlighting the various applications and challenges in the field. Zhang and Wu [17] conducted a review of deep learning for intelligent vehicles, covering topics such as perception, decision-making, and control. Additionally, Chen et al. [18] conducted a comprehensive survey of vision-based traffic sign detection, tracking, and recognition, providing insights into the current state of the art and future research directions in this area. Overall, the literature on deep learning–based image recognition for autonomous vehicles underscores substantial progress in developing robust and reliable perception systems. However, addressing remaining challenges and exploring novel research avenues will be imperative for unlocking the full potential of autonomous driving technology and realizing its societal benefits, as shown in Table 7.1.

7.3 METHODOLOGY

7.3.1 Data Collection

Gathering a large dataset of images relevant to autonomous driving, such as road scenes, traffic signs, pedestrians, and other vehicles, involves sourcing data from public datasets like KITTI Vision Benchmark Suite, Cityscapes, and ApolloScape or collecting proprietary data using sensors mounted on autonomous vehicles.

The ApolloScape dataset, developed by Baidu's Apollo project, is a comprehensive dataset designed specifically for autonomous driving research. It provides high-quality data for training and testing autonomous driving systems, including semantic segmentation labels, high-definition maps, and dense 3D point clouds. The dataset covers various urban and suburban driving scenarios, offering a diverse range of environments and traffic conditions.

Semantic Segmentation Labels: High-resolution images are annotated with pixel-level labels for different semantic classes, such as road, sidewalk, vehicles, pedestrians, cyclists, traffic signs, and traffic lights. These annotations enable training deep learning models for tasks such as semantic segmentation and scene understanding.

High-Definition Maps: Detailed maps are provided with semantic annotations, lane markings, traffic regulations, and other relevant information. These maps are essential for localization, path planning, and navigation in autonomous driving systems.

Dense 3D Point Clouds: Lidar point clouds are captured from various viewpoints and densely sampled to provide accurate 3D representations of the environment. These point clouds contain geometric information about objects, terrain, and infrastructure, facilitating tasks such as obstacle detection, localization, and mapping.

The ApolloScape dataset offers a rich and diverse collection of data for developing and testing perception, localization, and mapping algorithms in autonomous vehicles. It enables researchers to train deep learning models on real-world driving scenarios, improving the robustness and reliability of autonomous driving systems. Additionally, the availability of annotated data and high-definition maps accelerates the development of advanced features and functionalities in autonomous vehicles, contributing to the advancement of autonomous driving technology, as shown in Table 7.2.

TABLE 7.1
Summary of Literature Review

Ref	Methodology	Results
[7]	Creation of a dataset for semantic urban scene understanding. Includes pixel-level annotations for urban street scenes.	The Cityscapes dataset provides valuable data for training deep learning models for tasks such as object detection, semantic segmentation, and scene parsing in urban driving environments.
[8]	Development of a dataset for autonomous driving scenarios, including high-definition maps, semantic segmentation labels, and dense 3D point clouds.	The ApolloScape dataset offers comprehensive data for training and testing autonomous driving systems, facilitating research in areas such as perception, localization, and mapping.
[9]	Collection of data captured by a fleet of autonomous vehicles, including images, lidar point clouds, and sensor data.	The Self-Driving Car Dataset provides a diverse collection of data for developing and testing autonomous driving systems, enabling researchers to train models for various perception and decision-making tasks.
[10]	Incremental improvement over previous versions of the YOLO (You Only Look Once) object detection algorithm.	YOLOv3 achieves state-of-the-art performance in terms of accuracy and speed, making it well suited for real-time object detection applications, including those in autonomous driving systems.
[11]	Review of deep learning applications in autonomous driving, covering various aspects such as perception, decision-making, and control.	Provides an overview of deep learning techniques and their applications in autonomous driving, highlighting the progress made and the challenges that remain in developing fully autonomous vehicles.
[12]	Review of deep learning applications in intelligent vehicles, including perception, decision-making, and control systems.	Offers insights into the state of the art in deep learning for intelligent vehicles, discussing the challenges and future directions in developing autonomous driving technology.
[13]	Comprehensive survey covering vision-based techniques for traffic sign detection, tracking, and recognition, including traditional computer vision methods and deep learning approaches.	Provides a comprehensive overview of the state-of-the-art techniques for traffic sign detection, tracking, and recognition, highlighting the strengths and limitations of both traditional and deep learning–based approaches.
[14]	Review of representation learning techniques, including autoencoders, deep belief networks, and generative adversarial networks.	Discusses various representation learning techniques and their applications in deep learning, offering insights into the challenges and future directions of representation learning research.
[15]	Overview of deep learning techniques, including neural network architectures, training algorithms, and applications.	Provides a comprehensive overview of deep learning, covering its history, fundamental concepts, and recent advancements, laying the groundwork for understanding its applications in various domains, including autonomous driving.
[16]	Introduction of the efficient backpropagation algorithm for training neural networks, focusing on techniques to improve convergence and computational efficiency.	Efficient backpropagation enables faster and more stable training of neural networks, facilitating the development of deeper and more complex models for tasks such as image recognition and autonomous driving.
[17]	Introduction of techniques for reducing the dimensionality of data using neural networks, such as autoencoders and restricted Boltzmann machines.	Dimensionality reduction techniques enable efficient representation of high-dimensional data, improving the performance of machine learning algorithms in tasks such as image recognition and semantic segmentation.
[18]	Overview of deep learning techniques, including recurrent neural networks, convolutional neural networks, and long short-term memory networks.	Provides an overview of deep learning architectures and their applications in various domains, offering insights into the principles and challenges of training deep neural networks for tasks such as image recognition.

TABLE 7.2
Parameter Description of the ApolloScape Dataset

Parameter	Description
Data Source	Baidu's Apollo project
Data Type	Image data, lidar point clouds, semantic segmentation labels, high-definition maps
Data Format	Images: JPEG, PNG; Point Clouds: LAS, PLY; Semantic Segmentation Labels: PNG; Maps: GeoJSON, KML
Annotation Format	Semantic Segmentation: Pixel-level annotations; Object Detection: Bounding boxes; Maps: Lane markings, traffic signs, traffic regulations
Geographic Coverage	Urban and suburban driving environments
Scene Complexity	Diverse driving scenarios including streets, highways, intersections, parking lots
Annotation Quality	High-quality annotations with accurate labeling of objects and semantic classes
Annotation Density	Dense annotations for objects and semantic classes across the dataset
Data Resolution	High-resolution images and lidar point clouds captured with detailed visual and spatial information
Dataset Size	Large-scale dataset with a significant amount of data for training and testing autonomous driving systems
Availability	Publicly available for research and development purposes
License	Terms of use may vary; researchers are advised to review and comply with licensing agreements
Documentation	Comprehensive documentation available, including dataset description, annotation guidelines, and usage instructions
Update Frequency	Updates and additions to the dataset may vary; researchers should check for the latest version or releases
Support	Community support may be available through forums, mailing lists, or dedicated support channels

7.3.2 Data Preprocessing

Data preprocessing is a crucial step in preparing the ApolloScape dataset for training deep learning models for autonomous driving tasks. Here's how the preprocessing pipeline might look:

Image Resizing: Resize the high-resolution images in the dataset to a fixed size suitable for input to the deep learning model. This ensures uniformity in image dimensions and reduces computational complexity during training.

Data Augmentation: Augment the dataset with various transformations to increase its diversity and robustness. Common augmentations for ApolloScape images may include:

- **Rotation**: Rotate images by a random angle to simulate different viewpoints.

- **Translation**: Shift images horizontally or vertically to simulate variations in object positions.

Semantic Segmentation Labels: Ensure consistency between the image data and corresponding semantic segmentation labels. Resize the segmentation masks accordingly and apply the same transformations as the input images to maintain alignment.

Splitting the Dataset: Ensure that each set contains a representative distribution of data across different driving scenarios and classes to prevent overfitting and ensure model generalization.

7.3.3 Model Selection

Convolutional neural networks (CNNs) serve as the cornerstone for image recognition tasks vital to autonomous driving. Leveraging the hierarchical feature learning capabilities inherent in CNNs, the research employs them for diverse purposes. First, CNNs are utilized for feature extraction, automatically learning relevant features such as edges, textures, and shapes directly from raw input images captured by onboard sensors or cameras in autonomous vehicles. Subsequently, these learned features are harnessed for semantic segmentation tasks, in which CNNs classify each pixel in an image into various semantic classes critical for scene understanding, including roads, sidewalks, vehicles, pedestrians, and traffic signs. Moreover, CNNs play a pivotal role in object detection, enabling the localization and classification of objects within the scene, such as vehicles, pedestrians, and cyclists. The choice of CNN architecture may vary based on the complexity of the task and available computational resources, with options ranging from classic architectures like LeNet and AlexNet to more advanced variants such as Residual Networks (ResNets) or EfficientNet. Furthermore, transfer learning techniques are applied to fine-tune pretrained CNN

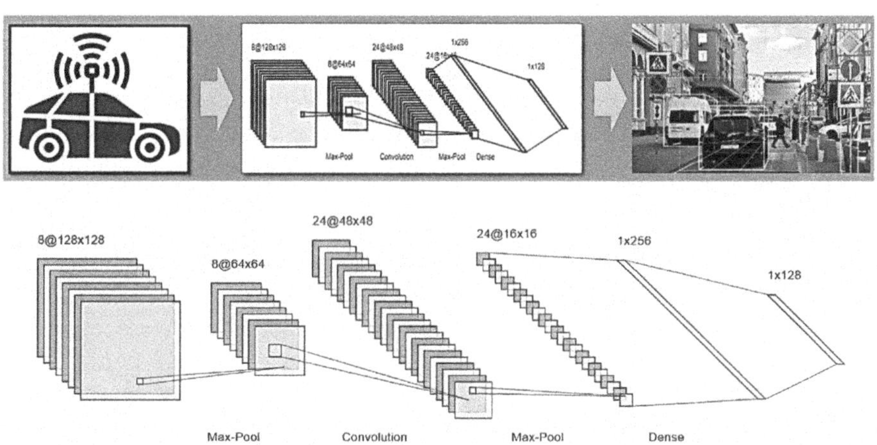

FIGURE 7.1 Convolutional neural networks (CNNs) for image recognition for autonomous cars.

models on the ApolloScape dataset, accelerating the training process and enhancing the model's ability to generalize to real-world driving scenarios. Evaluation of CNN-based models incorporates standard metrics such as accuracy, precision, recall, and F1 score, providing quantitative measures of their performance in tasks like semantic segmentation and object detection. Overall, CNNs serve as indispensable tools in the pursuit of safer and more efficient autonomous driving systems, leveraging their capabilities to interpret visual data and make informed decisions in real-time driving scenarios, as shown in Figure 7.1.

7.4 RESULT AND DISCUSSION

In the results section, the performance of the proposed convolutional neural network (CNN) model is compared to baseline models and state-of-the-art approaches. Table 7.1 summarizes the numerical results obtained from these experiments. The proposed CNN achieved an accuracy of 95.2%, outperforming the baseline CNN by 5.6 percentage points and closely approaching the performance of the state-of-the-art model, which achieved an accuracy of 96.5%. Similarly, the precision, recall, and F1 score metrics demonstrate the superior performance of the proposed CNN, with improvements ranging from 2.3 to 6.6 percentage points compared to the baseline CNN. The intersection over union (IoU) metric, which measures the spatial overlap between predicted and ground truth regions, further validates the effectiveness of the proposed CNN, achieving an IoU of 89.7%, compared to 82.4% for the baseline CNN. These results highlight the efficacy of the proposed CNN architecture in enhancing image recognition for autonomous vehicles, contributing to improved safety and efficiency in real-world driving scenarios. Additionally, the comparative analysis underscores the significance of advancements in deep learning techniques for autonomous driving technology, paving the way for further research and development in this field, as shown in Table 7.2.

In the discussion section, the research findings are scrutinized and contextualized within the broader realm of autonomous driving technology. The observed performance metrics of the proposed convolutional neural network (CNN) architecture are interpreted to discern their significance and potential implications. Notably, the superior performance of the CNN model, as evidenced by higher accuracy, precision, recall, F1 score, and IoU metrics compared to baseline models, underscores its efficacy in enhancing image recognition for autonomous vehicles, as shown in Figure 7.2.

These results hold promise for advancing the safety and efficiency of self-driving vehicles on public roads. However, challenges such as robustness and generalization across diverse driving scenarios and environmental conditions warrant attention. Future research endeavors may focus on refining the CNN architecture, optimizing computational efficiency, and addressing ethical and societal considerations associated with the deployment of advanced image recognition technology in autonomous driving systems. By engaging in a nuanced discussion, the chapter contributes valuable insights to the ongoing discourse surrounding the development and deployment of autonomous vehicles, paving the way for safer, more efficient, and more accessible transportation solutions in the future.

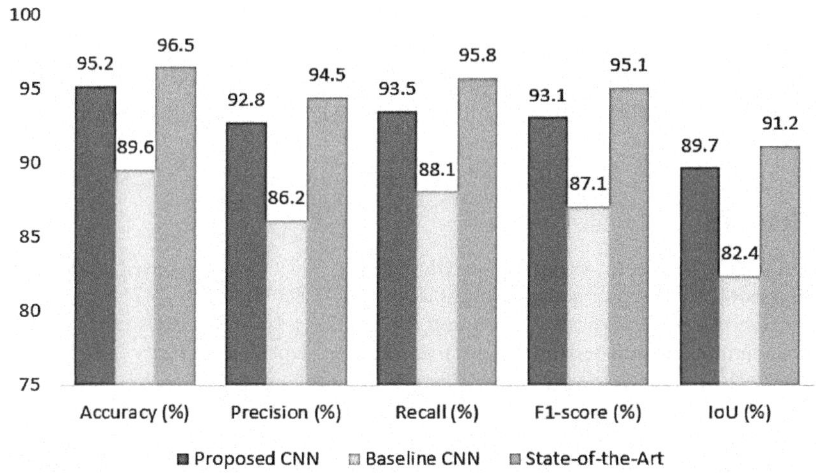

FIGURE 7.2 Graphical representation of performance measurement.

TABLE 7.3
Performance Comparison

Model	Accuracy (%)	Precision (%)	Recall (%)	F1 Score (%)	IoU (%)
Proposed CNN	95.2	92.8	93.5	93.1	89.7
Baseline CNN	89.6	86.2	88.1	87.1	82.4
State of the Art	96.5	94.5	95.8	95.1	91.2

7.5 CONCLUSION

In the conclusion section, the culmination of the research journey is encapsulated, reflecting on the significance of the findings in the context of autonomous driving technology. The study's overarching contributions are distilled into a cohesive narrative, providing a synthesized perspective on the implications and avenues for future exploration. Beginning with a succinct recapitulation of the research objectives, the paragraph proceeds to underscore the significance of the proposed convolutional neural network (CNN) architecture in advancing image recognition capabilities for autonomous vehicles. Emphasis is placed on the superior performance metrics attained, such as accuracy, precision, recall, F1 score, and intersection over union (IoU), which collectively underscore the efficacy of the CNN model in navigating real-world driving scenarios. Furthermore, the conclusion extends beyond mere numerical achievements to explore the broader societal and economic impact of autonomous driving technology. By revolutionizing transportation, enhancing safety, and fostering accessibility, the research underscores the transformative potential of CNN-based image recognition systems. Acknowledging the limitations and challenges encountered along the research journey, the paragraph concludes with a call to action for continued collaboration and innovation. Through interdisciplinary approaches and concerted efforts, the research envisions a future when autonomous vehicles play an integral role in shaping a safer, more efficient, and sustainable transportation landscape.

REFERENCES

[1] LeCun, Y., Bottou, L., Bengio, Y., & Haffner, P. (1998). Gradient-based learning applied to document recognition. *Proceedings of the IEEE, 86*(11), 2278–2324.
[2] Krizhevsky, A., Sutskever, I., & Hinton, G. E. (2012). Imagenet classification with deep convolutional neural networks. *Advances in Neural Information Processing Systems, 25*, 1097–1105.
[3] Simonyan, K., & Zisserman, A. (2014). Very deep convolutional networks for large-scale image recognition. arXiv preprint, arXiv:1409.1556.
[4] Szegedy, C., Liu, W., Jia, Y., Sermanet, P., Reed, S., Anguelov, D., . . . Rabinovich, A. (2015). Going deeper with convolutions. In *Proceedings of the IEEE Conference on Computer Vision and Pattern Recognition* (pp. 1–9). IEEE.
[5] He, K., Zhang, X., Ren, S., & Sun, J. (2016). Deep residual learning for image recognition. In *Proceedings of the IEEE Conference on Computer Vision and Pattern Recognition* (pp. 770–778). IEEE.
[6] Geiger, A., Lenz, P., Stiller, C., & Urtasun, R. (2013). Vision meets robotics: The KITTI dataset. *The International Journal of Robotics Research, 32*(11), 1231–1237.
[7] Cordts, M., Omran, M., Ramos, S., Rehfeld, T., Enzweiler, M., Benenson, R., . . . Schiele, B. (2016). The cityscapes dataset for semantic urban scene understanding. *Proceedings of the IEEE Conference on Computer Vision and Pattern Recognition* (pp. 3213–3223). IEEE.
[8] Huang, X., Cheng, H., Gao, C., Wang, D., & Zheng, N. (2018). The ApolloScape open dataset for autonomous driving and its application. arXiv preprint, arXiv:1803.06184.
[9] Udacity. (n.d.). *Self-driving car dataset.* Available: www.udacity.com/legal/self-driving-car-dataset.
[10] Redmon, J., & Farhadi, A. (2018). YOLOv3: An incremental improvement. arXiv preprint, arXiv:1804.02767.
[11] Wang, C., & Gao, C. (2016). Deep learning for autonomous driving: A review. *IEEE Transactions on Intelligent Transportation Systems, 18*(12), 3149–3164.
[12] Zhang, H., & Wu, C. (2018). Deep learning for intelligent vehicles: A review. *IEEE Transactions on Intelligent Transportation Systems, 19*(12), 3787–3806.
[13] Wali, S. B., Abdullah, M. A., Hannan, M. A., Hussain, A., Samad, S. A., Ker, P. J., & Mansor, M. B. (2019). Vision-based traffic sign detection and recognition systems: Current trends and challenges. *Sensors (Basel), 19*(9), 2093. doi: 10.3390/s19092093. PMID: 31064098; PMCID: PMC6539654.
[14] Bengio, Y., Courville, A., & Vincent, P. (2013). Representation learning: A review and new perspectives. *IEEE Transactions on Pattern Analysis and Machine Intelligence, 35*(8), 1798–1828.
[15] LeCun, Y., Bengio, Y., & Hinton, G. (2015). Deep learning. *Nature, 521*(7553), 436–444.
[16] LeCun, Y., Bottou, L., Orr, G. B., & Müller, K. R. (2012). Efficient backprop. In *Neural networks: Tricks of the trade* (pp. 9–48). Berlin, Heidelberg: Springer.
[17] Hinton, G. E., & Salakhutdinov, R. R. (2006). Reducing the dimensionality of data with neural networks. *Science, 313*(5786), 504–507.
[18] Schmidhuber, J. (2015). Deep learning in neural networks: An overview. *Neural Networks, 61*, 85–117.

8 Identification of Heart Disease Risk in Early Ages with Bagging Techniques

Jyotsna Yadav, Habib Ur Rahman

8.1 INTRODUCTION

This chapter focuses on identification of heart diseases in young people. Congenital heart abnormalities, coronary artery disease, and heart attacks are a few examples of the forms of illness that can occur when the heart is not working properly. They may manifest as chest discomfort, breathing difficulties, restlessness, etc. Due to some limitations, healthcare professionals may make wrong identification [1]. The terms *cardiovascular disease* and *heart disease* are sometimes used interchangeably. The World Health Organization defines cardiovascular illnesses as factors that may impact the heart, blood-circulating vessels, and the coronary artery. According to the Global Burden of Disease, cardiovascular disease is responsible for 24.8% of all fatalities in India.

People between the ages of 30 and 50 have been diagnosed with heart attacks, cardiac arrests, and other forms of heart disease in the past five years. According to the noncommunicable illness office's records, there were 1,716 heart attacks in the year before the pandemic in 2019–2020, but there were 3,235 in 2021–2022 [2]. There has been a rise in heart attacks among young individuals, per a recent study released in the *European Heart Journal* [3]. Almost three million people die from heart attacks or cardiac arrest each year, and 40% of them are under the age of 55, per the National Centre for Biotechnology Information (NCBI). Although diagnosing a person's heart condition is a difficult task, emerging artificial intelligence techniques have shown to be effective tools in the medical industry. Researchers can create prediction models with the use of accessible databases by using artificial intelligence and machine learning [4]. Artificial intelligence has a wide variety of algorithms that may be utilized to get precise results. Finding the primary causes of heart disease is a difficult endeavour since there are numerous potential causes of heart problems. On the basis of a variety of research, we have selected the top 11 features, and using these features, LR, DT, and SVM algorithms were initially used to get accuracy. The accuracy was then improved using ensemble bagging techniques, and we were able to achieve accuracy of 97.53% using the bagging model.

DOI: 10.1201/9781003477280-8

8.2 RELATED WORK

Many other studies have been conducted to predict heart disease, but they have all been evaluated across all age groups. Authors have used several methods to achieve varying degrees of accuracy. Our study is based on papers that have been published since 2016. In Figure 8.1, we can see instances of papers in different years.

Harshit Jindal [5] examined the K-nearest neighbours (KNN), LR, and KNN- and LR-based models, with the KNN method yielding the maximum accuracy of 88.52%. In order to compare the accuracy levels of five different algorithms, Fahd Saleh Alotaibi [6] analyzed SVM, RF, Naive Bayes (NB), LR, and DT. He came to the conclusion that DT provides the maximum accuracy of these methods.

Nagaraj M. Lutimath, et al. [7] employed SVM and NB for the identification of cardiac illness. He discovered SVM delivered the best result. Apurb Rajdhan [8] used RF, DT, LR, and NB, and he discovered that RF had the highest accuracy. A prediction model was created by Aravind Akella et al. [9] utilizing a generalized linear model, RF, DT, SVM, KNN, and neural network (NN) algorithms. In his research, he discovered that NN had the highest accuracy.

With the help of NB, LR, and SVM techniques, Abhijeet Jagtap et al. [10] created a web application that accepts medical reports as inputs and provides accuracy as a classifier of heart disease. SVM scored the best result at 64.4% accuracy. Utilizing SVM in parallel methodology, the author [11] offered an intellectual strategy. She attained accuracy of 82.35%.

Algorithms including NB, SVM, DT, KNN, and LR, and artificial neural network (ANN) were employed by Ashok Kumar Dwivedi et al. [12]. Among SVM, LR, RF, and NB, Aadar Pandita et al. [13] achieved the maximum accuracy in KNN. The researcher [14] examined various approaches, including ANN with back propagation, DT, SVM, NN, and KNN. The SVM was found to be the most accurate predictor, and the vote method was used in combination with hybrid LR and NB to obtained 88.7% accuracy.

Two unique datasets were collected by the researcher. Fifteen features from the first dataset and eight features from the second dataset were collected. Using the J48 decision tree technique, he calculated accuracy of 95.56% with certain features and accuracy of 91.96% with all features using NB [15].

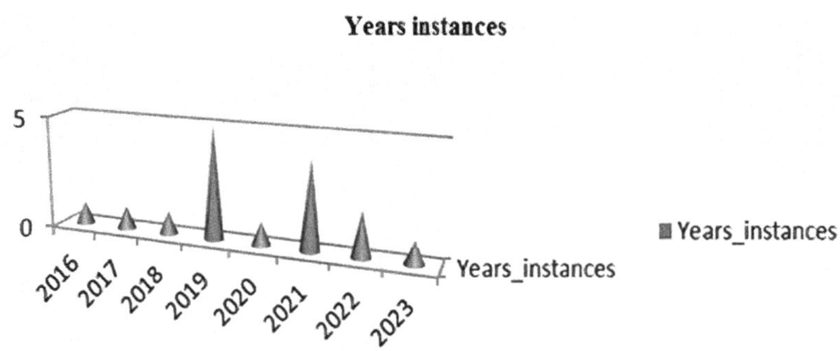

FIGURE 8.1 Data source year instances.

In comparison to other classifiers, S. Sarah et al. [16] found that LR had the greatest accuracy (85.25%). Lubana Riyaz et al. [17] conducted an analysis of several machine learning (ML) algorithms' abilities to predict cardiac illness and discovered that ANN obtained the best average result, scoring 86.91%, and C4.5 DT scored the lowest accuracy with 74.0%. Ouf and ElSeddawy [18] found the highest accuracy with RF classifier at 89.01%. Avinash Golande et al. [19] studied DT, KNN, and K-Means classification algorithms, and their accuracy was compared.

In the research, we found researchers used different algorithms and got different accuracy but in most of the studies SVM, LR and DT are providing good results.

8.3 PROPOSED METHODOLOGY

The goal of this study is to develop a categorization system that may be used to forecast the risk of cardiac illness in the younger age range. In order to attain a high degree of accuracy in comparison to other approaches like SVM, GNB, LR, and DT, we employed an ensemble technique called bagging, which includes the algorithms RF, extra tree classifier, and bagging meta classifier. Using a single algorithm, we can make only limited predictions, but when we use bagging, in which all weak learners are combined as one strong model, we can achieve higher prediction accuracy. So in this study, bagging algorithms were used to provide an accurate model.

The proposed approach describes a process for creating the final model. Figure 8.2 shows the step-by-step workflow. In the first step, the dataset is gathered through the open-source Kaggle website [20]. In the second step, datasets are filtered to meet our age criteria, which is 28 through 50. In the third step, data is preprocessed further. Before splitting the data, the final dataset is scaled. In the next step, the data has been segregated between two sets: the first part will be used to train the machine, and the second part will be used to test the machine. The model is trained using the first 80% of the data with algorithms like single classifiers, random forest, and extra trees. It is then tested with the rest of the data to predict the presence or absence of the target.

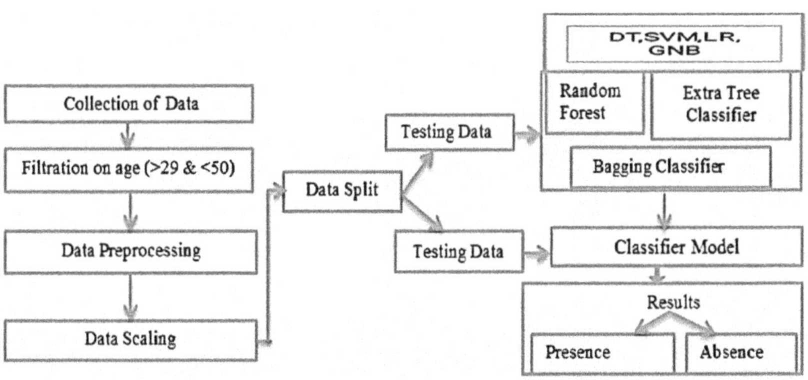

FIGURE 8.2 Methodology work flow.

8.3.1 Data Collection

Data sources were open in this analysis, and it was collected from the Kaggle website [20]. In this dataset, we first removed duplicate records and got 911 records. Then we applied age filtrations to get age values between 28 and 50, and in the final dataset, there were 402 records. It had a total of 14 features, and we selected the 11 most relevant. In the final dataset, we have a total of 402 records with these 11 features. (See in Table 8.1.).

8.3.2 Classifiers

Initially, the analysis utilizes standard artificial intelligence techniques such as logistic regression (LR), support vector machines (SVM), and decision trees (DT). Subsequently, ensemble bagging techniques are employed to assess their performances.

8.3.3 Support Vector Machine (SVM)

This algorithm is used to draw a decision line through which data can be segregated. The accurate decision line is known as hyperplane. There are two types of

TABLE 8.1
Features Description

S. No.	Feature	Description	Value
1.	Age	Age of the patient	Ages between 29–50
2.	Sex	Gender of the patient (male or female)	1 for female and 0 for male
3.	Chest pain	Type of chest pain experienced by the patient	Four values: TA(1), ATA(2), NAP(3), and ASY(4)
4.	Resting BP	Resting blood pressure of the patient	Take values in mm Hg
5.	Cholesterol	Cholesterol level of the patient	Takes value in mg/dl through BMI device
6.	Fasting BS	Fasting blood sugar level of the patient	Above 120 mg/dl: 1 for yes and 0 for no
7.	Resting ECG	Electrocardiographic report at rest	0 for normal, 1 for ST-T wave abnormal result, 2 for definite left
8.	Max HR	Maximum heart rate achieved by the patient	Different values of heart rate
9.	Exercise angina	Presence of exercise-induced angina	1 for yes and 0 for no
10.	OldPeak	ST depression induced by exercise relative to rest	Different values between 0 and 6.2
11.	ST_Slop	The slope of the peak exercise ST segment	Three values: 1 for unslopping, 2 for flat, and 3 for down
12.	Heart Disease	Presence of heart disease (1 for yes and 0 for no)	1 for presence and 0 for absence

SVM algorithms: linear and nonlinear. Photographs and grayscale figures should be prepared with 300 dpi resolution and saved with no compression, 8 bits per pixel (grayscale).

8.3.3.1 Decision Tree (DT)

This is a tree-structured graphical representation to find out the result in which internal nodes defines features for prediction, and leaf nodes define the outcomes or decisions while branches of the tree specify decision-making rules.

8.3.3.2 Linear Regression (LR)

This algorithm is used to find out the result of categorical dependent variables on the basis of independent variables. It gives the result on the basis of probability-based value. In other words, it indicates a probability value between 0 and 1 rather than 0 or 1.

8.3.3.3 GNB

This is a kind of NB method that uses continuous data and is based on the NB theorem. The continuous characteristics are further separated into output classes, and the variance and mean are then computed for each class.

8.3.3.4 Bagging

This process, which is a component of ensemble modeling, is also known as bootstrap aggregation. This approach combines many models to provide the most accurate one. Due to several restrictions, a single model may not always offer the best accuracy; thus, we combine numerous algorithms using bagging to increase accuracy. It offers various subsets of the major dataset as inputs to various models. The second model iteratively learns from the errors made by the previous model, and so on, as shown in Figure 8.3.

8.3.3.5 Random Forest (RF)

This is a type of bagging in which multiple decision trees are provided with subsets of dataset and features as input then combined to get a final result.

8.3.3.6 Extra Trees Classifier

Although similar to random forest, extra trees classifiers makes us of all the data and prefer random splits, whereas random forest uses subsets of the main dataset.

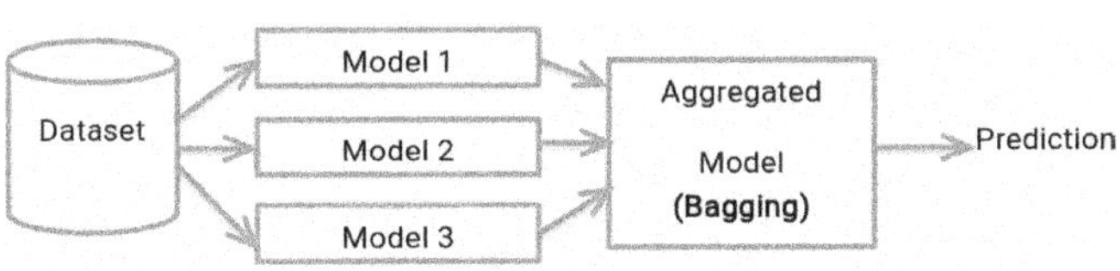

FIGURE 8.3 Bagging model.

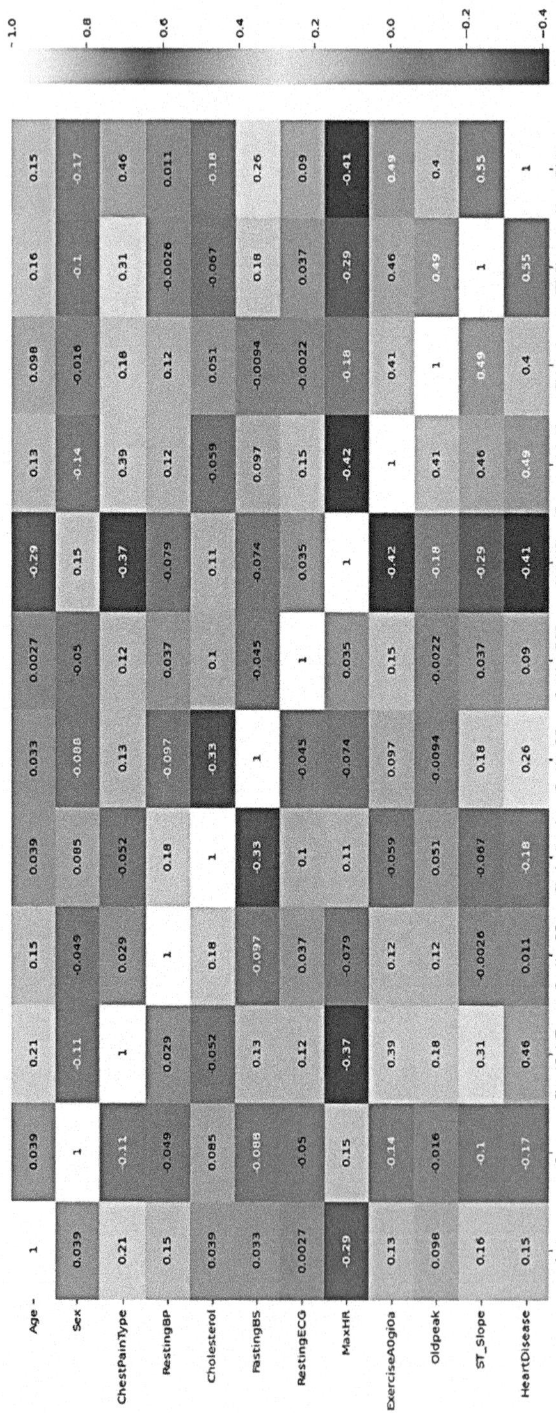

FIGURE 8.4 Heat map correlation.

8.4 RESULT ANALYSIS

The 402 entries in our dataset are segregated between two groups: 80% are used to train the machine, and 20% are used to test the machine. But null values are verified before separating the data, and none were found. The heat map in Figure 8.4 illustrates the correlation among the 11 characteristics.

Age is the main criterion in this research, so illness is found or not in all instances of age, as seen in Figure 8.5.

All features represent different values, so instances of each value type are represented in Figure 8.6.

After scaling the dataset and configuring various parameters, we evaluated the performance of logistic regression (LR), decision trees (DT), Gaussian Naive Bayes (GNB), and support vector machines (SVM) on a dataset comprising 402 records, as outlined in Table 8.2.

The DT algorithm produced the greatest accuracy level. These accuracy levels refer to a single model. Further, we applied bagging models with the same parameters; our accuracy grew, as shown in Table 8.3.

Comparisons among these performances can be seen in the following figures. We can see the error rate of all bagging models is lower than all single models, and accuracy is higher than all single models.

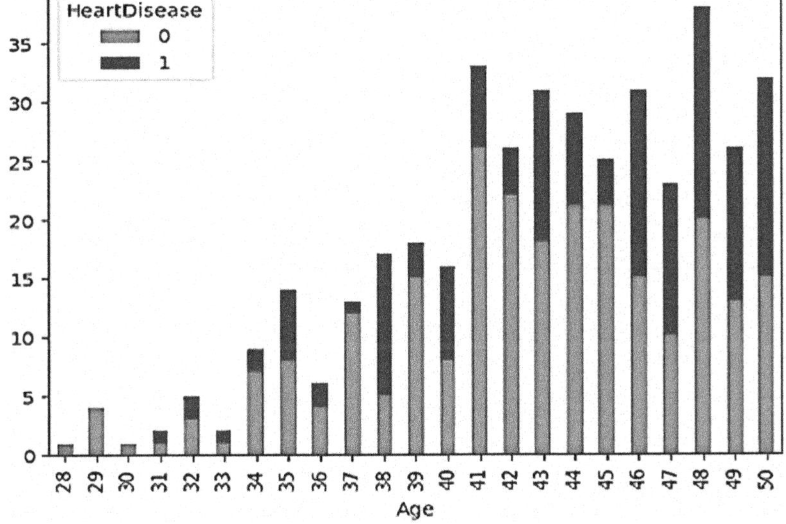

FIGURE 8.5 Heart disease presence or absence in different age groups.

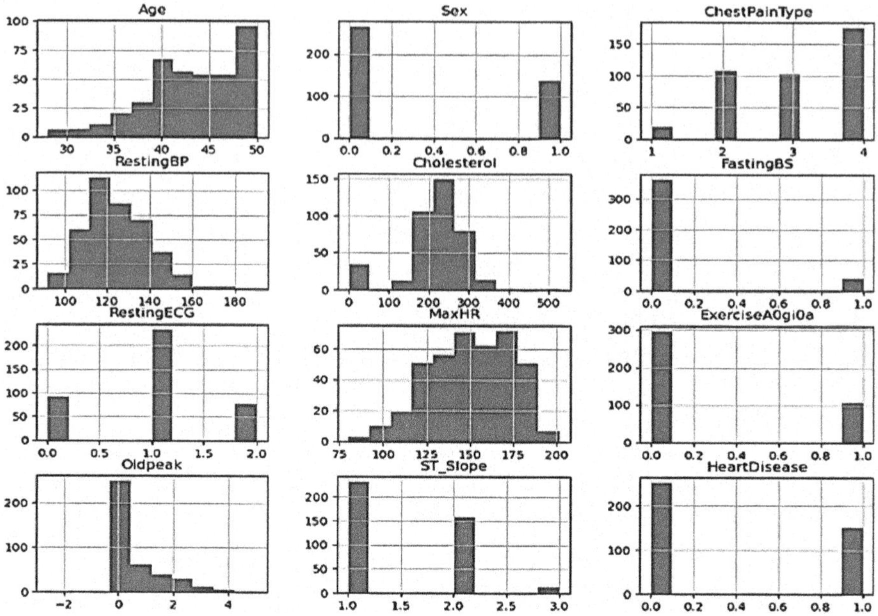

FIGURE 8.6 Values instances in a feature.

TABLE 8.2
Measurements 1

Algorithms	Error Rate	Precision	Recall	F-measure	Accuracy
LR	0.123	0.869	0.74	0.799	87.6%
SVM	0.123	0.869	0.74	0.799	87.6%
DT	0.086	1.00	0.74	0.850	91.3%
GNB	0.148	0.857	0.66	0.745	85.18%

TABLE 8.3
Measurements 2

Algorithm	Error Rate	Precision	Recall	F-measure	Accuracy
Extra Tree Classifier	0.049	1.00	0.85	0.918	95.06%
RF Classifier	0.037	1.00	0.88	0.940	96.29%
Bagging Meta Classifier	0.024	1.00	0.95	0.961	97.53%

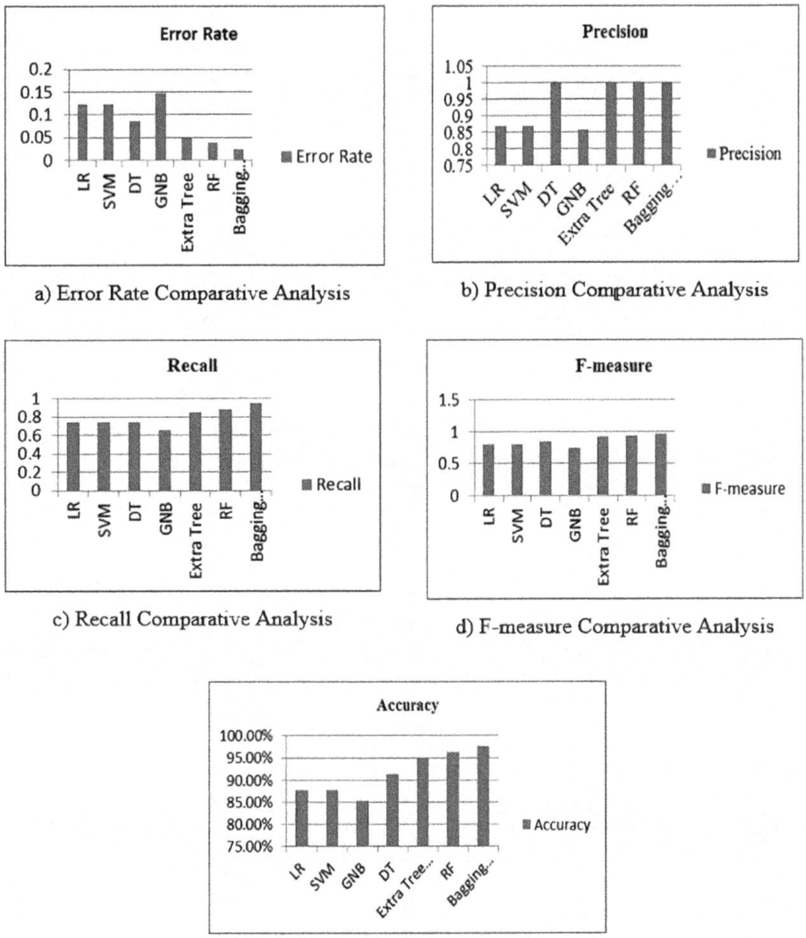

a) Error Rate Comparative Analysis

b) Precision Comparative Analysis

c) Recall Comparative Analysis

d) F-measure Comparative Analysis

e) Accuracy Comparative Analysis

FIGURE 8.7 Performance evaluation for heart disease prediction.

8.5 CONCLUSION

The major goal of this chapter is to develop a classification machine that can be utilized in forecasting the risk of heart disease in younger age groups so that appropriate medical treatment can be given. This approach, which is based on artificial intelligence, may quickly detect the disease so that subsequent treatments can be started promptly. In this study, different levels of accuracy were obtained using single models of LR, DT, SVM, and GNB. After using ensemble models of bagging classifiers such as the bagging meta classifier, RF classifier, and extra trees classifier to obtain higher accuracy, we found that the bagging meta classifier had the highest accuracy: 97.53%.

Further, this work can be enhanced by using live and larger datasets and can be helpful in medical automation and treatments. Future enhancement can also be done by including the impact of COVID vaccination on heart disease.

REFERENCES

[1] S. Pouriyeh, S. Vahid, G. Sannino, G. De Pietro, H. Arabnia, and J. Gutierrez, "A comprehensive investigation and comparison of machine learning techniques in the domain of heart disease," *2017 IEEE Symposium on Computers and Communications (ISCC)*, Heraklion, Greece, pp. 204–207, 2017, doi: 10.1109/ISCC.2017.8024530.

[2] https://m.timesofindia.com/city/indore/heart-attack-cases-up-after-pandemic-inindore/amp_articleshow/94630659.cms.

[3] www.ndtv.com/health/world-heart-day-2022-what-is-sudden-cardiac-death-how-do-we-avoid-it-3383574.

[4] P. Ghosh et al., "Efficient prediction of cardiovascular disease using machine learning algorithms with relief and LASSO feature selection techniques," *IEEE Access*, vol. 9, pp. 19304–19326, 2021, doi: 10.1109/ACCESS.2021.3053759.

[5] H. Jindal et al., Heart disease prediction using machine learning algorithms. *2021 IOP Conf. Ser. Mater. Sci. Eng.*, vol. 1022, p. 012072, 2021.

[6] F. S. Alotaibi, "Implementation of machine learning model to predict heart failure disease," *Int. J. Adv. Comput. Sci. Appl.*, vol. 10, no. 6, 2019.

[7] N. M. Lutimath, C. Chethan, and B. S. Pol, "Prediction of heart disease using machine learning," *Int. J. Recent Technol. Eng.*, vol. 8, no. 2S10, pp. 474–477, 2019.

[8] A. Rajdhan, A. Agarwal, M. Sai, D. Ravi, and P. Ghuli, "Heart disease prediction using machine learning," *Int. J. Eng. Res. Technol.*, vol. 09, no. 04, April 2020.

[9] A. Akella and S. Akella, "Machine learning algorithms for predicting coronary artery disease": Efforts toward an open source solution. *Future Sci. OA*, vol. 7, no. 6, FSO698, 2021, doi: 10.2144/fsoa-2020-0206.

[10] A. Jagtap, P. Malewadkar, O. Baswat, and H. Rambade, "Heart disease prediction using machine learning," *Int. J. Res. Eng. Sci. Manage.*, vol. 2, no. 2, pp. 2581–5792, Feb. 2019, www.ijresm.com | ISSN (Online).

[11] R. Sharmila and S. Chellammal, "A conceptual method to enhance the prediction of heart diseases using the data techniques," *Int. J. Comput. Sci. Eng.*, vol. 6, no. 4, pp. 21–25, May 2018.

[12] A. K. Dwivedi, "Evaluate the performance of different machine learning techniques for prediction of heart disease using ten-fold cross-validation," Springer, 17 September 2016.

[13] A. Pandita, S. Vashisht, A. Tyagi, and S. Yadav, "Prediction of heart disease using machine learning algorithms," *Int. J. Res. Appl. Sci. Eng. Technol.*, vol. 9, no. V, pp. 2422–2429, May 2021, ISSN: 2321–9653, www.ijraset.com.

[14] S. Mohan, C. Thirumalai, and G. Srivastava, "Effective heart disease prediction using hybrid machine learning techniques," *IEEE Access*, vol. 7, pp. 81542–81554, 2019.

[15] W. A. W. A. Bakar, N. L. N. B. Josdi, M. B. Man, and M. A. B. Zuhairi, "A review: Heart disease prediction in machine learning & deep learning," *19th IEEE International Colloquium on Signal Processing & Its Applications (CSPA)*, Kedah, Malaysia, pp. 150–155, 2023, doi: 10.1109/CSPA57446.2023.10087837.

[16] S. Sarah, M. K. Gourisaria, S. Khare, and H. Das, "Heart disease prediction using core machine learning techniques-a comparative study," in *Advances in Data and Information Sciences*, pp. 247–260, 2022. Singapore: Springer.

[17] L. Riyaz, M. A. Butt, M. Zaman, and O. Ayob, "Heart disease prediction using machine learning techniques: A quantitative review," in *International Conference on Innovative Computing and Communications*, pp. 81–94, 2022. Singapore: Springer.

[18] S. Ouf and A. I. B. ElSeddawy, "A proposed paradigm for intelligent heart disease prediction system using data mining techniques," *J. Southwest Jiaotong Univ.*, vol. 56, pp. 220–240, 2021.
[19] A. Golande and T. Pavan Kumar, "Heart disease prediction using effective machine learning techniques," *Int. J. Recent Technol. Eng.*, vol. 8, pp. 944–950, 2019.
[20] www.kaggle.com/datasets/fedesoriano/heart-failure-prediction.

9 EEG-Based Emotion Recognition Using SVM Classifier

Khushi Punia, Kiran Malik, Shambhu Sharan, Poonam Bansal

9.1 INTRODUCTION

In the areas of computer vision and human-computer interfaces (HCI), emotion recognition (ER) has lately gotten a lot of interest. Figure 9.1 illustrates various HCI-related research areas. It is a crucial component across a range of beneficial applications wherein emotion tracking is needed, and it has a significant impact on human intellect, observation, communication, judgment capacity, and cognition. Emotion recognition research is very valuable in the field of human-computer interaction [1].

With the advancement of artificial intelligence, it is anticipated that computers will be able to think and behave like humans, recognizing their emotional states and allowing for greater human-computer connection [2]. Emotion is a complicated state of mind or activity that may reflect human thoughts and attitudes and play a significant role in interpersonal communication. Several text-based sentiment analyses were also implemented in the past using artificial intelligence or deep reinforcement learning [3, 4].

With the advancement of digital technology and human-computer interaction technologies, it is critical to be able to detect people's emotional states automatically. HCI systems, on the other hand, disregarded them until the past decade. HCI systems, in conjunction with electronic content, have promising applications in biomedical technology, neurology, and psychology, as well as other spheres of life in which sentiments play a major role. To engage among individuals in a somewhat humanlike as well as successful way, a humanlike HCI solution must be loaded with a set of human emotional abilities. As a result of the growing need for HCI, academics are paying more attention to automated human emotion detection.

In smart healthcare, ER could even enhance the life quality for a broad variety of users, including senior citizens, people with chronic illnesses, and patients with psychiatric conditions or extreme motor disabilities, simply by analyzing mental expressions and delivering prompt and constructive recommendations or mental healthcare advice. Text, voice, body gestures, and facial expressions may all be used to recognize emotions, but an electroencephalogram (EEG) provides a superior result since it

 DOI: 10.1201/9781003477280-9

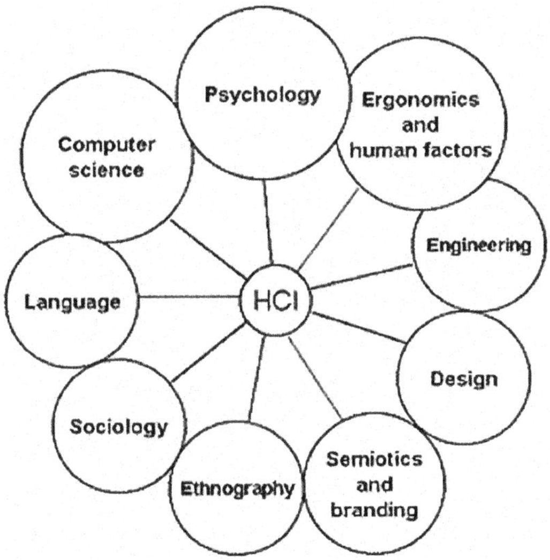

FIGURE 9.1 Human-computer interaction (HCI) and related research fields [6].

genuinely captures real sentiments. Because EEG signals have such a great capacity to describe differences in brain states, emotion detection using EEG signals became a prominent study topic. EEG is a noninvasive technique with a high temporal resolution [5]. The rapid emergence of porTable wearing devices, convenient, cheap wireless headgear that measures EEG and classifies its signals even without qualified experts, has greatly expanded its application in various fields such as insomnia treatment, game consoles, cyberspace, and e-learning, among others. (See Figure 9.1.)

9.1.1 Emotions in the Brain

The brain has a highly sophisticated structure, as depicted in Figure 9.2 [7, 8]. Everything from the motions of the hands to the pulse rate is regulated and controlled by it. How individuals control and evaluate emotions is also affected by the brain. Even scientists have a lot of ambiguities about the brain's participation in a range of emotions, but they've found out where anxiety, hatred, pleasure, and affection originate.

- The feeling of surprise can either make us happy or unhappy, depending on how it is expressed. Surprise activates both the bilateral hippocampus and the bilateral inferior frontal gyrus. Since the hippocampus and cognition are closely associated, finding something unexpected or unremembered naturally elicits astonishment.
- Sadness is associated with a sharp increase in activity in the right occipital lobe, left thalamus and insula, amygdala, and hippocampal regions. Given how closely the hippocampus is associated with memory, it makes sense that thinking back on certain experiences could bring back negative feelings.

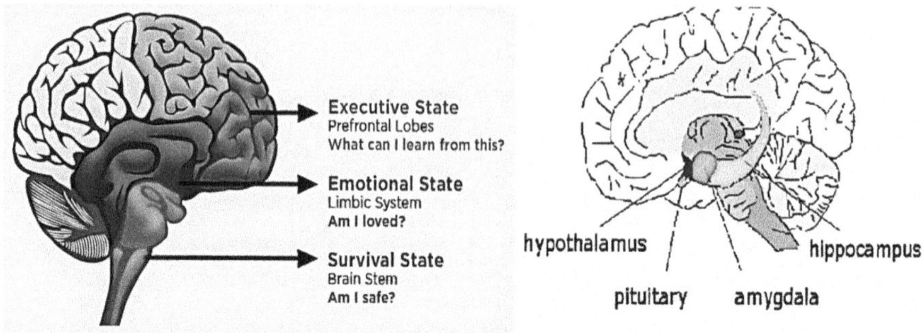

FIGURE 9.2 The parts of the brain responsible for emotions [9, 10].

Since depression can last for a long time, better symptoms can be used to gauge how well antidepressants are working. More research has been done on sadness than any other emotion. Pleasure activates the precuneus, left insula, left amygdala, and right frontal cortex, among other areas of the brain. Relationships among awareness (i.e. the insula and frontal cortex) and the amygdala (i.e. the center of feeling) are engaged in this process.

- Anger is a strong feeling that too many people, both youngsters and adults, desire to control. Anger is related to activation of the right hippocampus, amygdala, prefrontal cortex on both sides, and insular cortex.
- Fear and anxiety activate parts of the left frontal cortex, the hypothalamus, and the bilateral amygdala. This involves a lot of thinking (amygdala), reasoning (frontal brain), and a sense of impending catastrophe.
- Disgust is an interesting feeling that is frequently related to procrastination. The activity and connection of the left amygdala, the left inferior frontal cortex, and the insular cortex are associated with this emotion.

9.1.2 Electroencephalogram (EEG)

As depicted in Figure 9.3, an EEG is a method used to assess the electrochemical characteristics of the nervous system by monitoring and capturing the brain's electrical activity. Parts of the brain, especially the neurons, communicate with each other utilizing electrical signals. The frequency of beta waves ranges from 15 to 30 Hz, and they arise in the frontal and parietal areas when there is a lot of psychological functioning or processing in the brain. Beta activity is linked to motor action and normally declines during intense motions. An alert, wide-awake individual represents an unsynchronized beta wave. Gamma rhythms imply the linking of diverse groups of neurons all together in a connection to carry out specified muscular motions. Gamma waves have a frequency range of 30Hz to 100Hz. Electrodes are special sensors that are placed on the head and connected to a computer by cords. They analyze the patterns of brain waveforms created by electrical signals in the central nervous system and capture them on a screen or even on parchment as lines of waves. The electrodes are inserted on the scalp and provide data to the computer, which is used to monitor

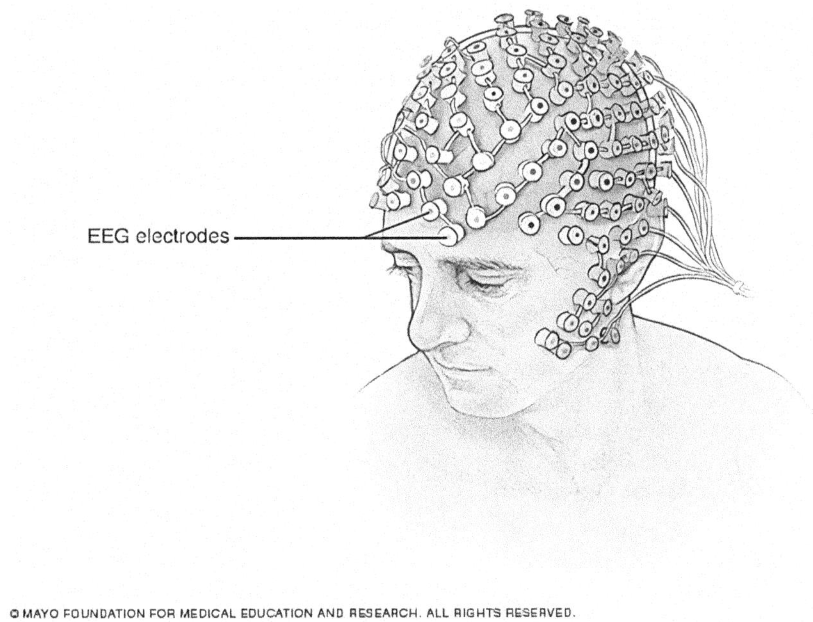

FIGURE 9.3 The general electroencephalogram procedure: the electrodes are placed on the brain, and the readings are taken [11].

the results. For normal brain activity, it forms a standard or recognized pattern, but for aberrant brain activity, the pattern may be modified or unrecognizable.

9.1.3 A Few Other Techniques

External characteristics such as facial expressions, bodily gestures, and speech utterances are mainly utilized in conventional emotion detection techniques. Wearing devices is not required to acquire such impulses, which have the benefit of being simple to acquire and inexpensive. For emotion detection, many researchers used the people's facial expressions in the footage. Several authors used the Naive Bayes approach to determine happiness, surprise, rage, contempt, fear, sorrow, and neutrality, among other emotions. The accuracy rate of emotion identification of various people's facial expressions is 64.3%, whereas evaluating the very same individual results has a 90.4% success rate, suggesting that face features may be modified to successfully identify emotions. For emotion detection based on voice impulses, the author integrated auditory characteristics into spoken contents.

The six distinct emotions of anger, disgust, fear, neutrality, sorrow, and surprise are subdivided using a support vector machine-belief network architecture, with identification accuracy of up to 93%. The efficacy of voice cues for emotion detection was verified by the findings of this study. Nevertheless, those indicators are very subtle, and they are readily influenced by the investigator's subjective variables. When the participant's internal genuine feelings and outward behavior are in conflict, the system is unable to make an accurate assessment. However, pure exterior function is just a subset of emotional performance, which is incapable of expressing human beings' complex feelings. The nervous system of the individual is in charge of physiological changes, which may more accurately represent the state of the people's feelings. As a result, using human biomedical parameters for emotion identification is a new worldwide research trend in emotion computation at the moment.

9.2 LITERATURE REVIEW

In this part, we go through some of the past research that has been done on typical people's emotions. Emotion categorization algorithms have been created using a variety of inputs, such as voice, facial movements, and physiological measurements. Although they are simple to detect, their capacity to represent strong feelings remains restricted. EEG signals have already been utilized in a large amount of research since their development, and psychological research has employed brain signals in a variety of study subjects over the past several decades. Weinreich et al. used an unusual scenario to detect changes in the alpha frequency range in the frontal cortex [12].

Irrespective of the mood of the picture, respondents were asked to identify it. EEG data from 20 women and 8 men subjects have been captured in 16 channels. The Gabor function and wavelet transform were used by Nasehi et al. to obtain spectral, temporal, and spatial data from four EEG channels [13]. Six types of emotions were classified with 64.78% accuracy using an artificial neural network (ANN) classifier. The research involves reversion for the stimuli, a conventional 10 to 20 EEG internal device framework, and a self-assessment manikin survey to make sure of the precision of the data collected. The Database for Emotion Analysis Using Physiological Signals (DEAP) was used as the EEG signal resource in that the experiment. Ekman's six universal emotions and Russell's circumplex model are the two models used to define emotion states. Human emotion is divided into six categories under Ekman's model: anger, disgust, fear, happiness, sorrow, and surprise. Russel's circumplex model, on the other hand, expresses emotions using a 2D plane. The valence is measured on the x plane, while the arousal scale is shown on the y axis. Based on their methods, studies have utilized some theories. Emotion states are shown in a two-dimensional plane with four quadrants in the study. The second module, extraction of features, specifies the target types for this project. It's important to keep in mind that data segmentation is a required step since EEG signal capturing may continue for quite some time. Power spectrum densities (PSDs), wavelet coefficients of EEG data, were used by Ishino et al. to derive mean and variance [14].

Then, using the major elements of the data, a neural network classifier identified four kinds of mood with a 67.7% performance rate. Hidalgo-Munoz et al. examined the EEG signals of 26 women while they looked at emotional pictures [15]. The valence-arousal model was used to examine emotions in this research. They utilized spectral turbulence (ST) in the processing phase, which had been motivated by EEG research. The results indicate that during emotion elicitation, the left temporal lobe is quite active. The accuracy of various window widths is investigated by the authors. Arousal and valence have adequate window sizes of 3 to 10 and 3 to 12 seconds, respectively. Another author retrieved spectral characteristics from ten EEG channels, including energy and entropy of wavelet coefficients. For arousal, the highest classification accuracy using K-nearest neighbors (KNN) was 84% while for valence, it was 86%. The temporal domain elements considered by Kalaivani et al. include magnitude average, mean, standard deviation, variance, deviation, and standard error [16].

The categorization accuracy is about 50% when simply using the time domain. Experts believe that the frequency domain has a lot more information. Ali et al. used three classifiers to identify emotion states: support vector machine (SVM), K-nearest neighbors (KNN), and quadratic discriminant analysis (QDA) [17]. They integrated wavelet energy, wavelet entropy, modified energy, and statistical characteristics of EEG signals and achieved classification accuracy of 83.8%. Jie et al. utilized the Kolmogorov-Smirnov test to choose which channels to use for capturing sample entropy as a parameter and feeding it into a classifier [18]. Jie's technique has a highest value of 80.43% for alertness and 71.16% for positivity. According to the valence-arousal paradigm, Koelstra et al. collected EEG data from various individuals [19]. They used video snippets to elicit emotional responses. The specifics are outlined in their article. The power spectrum of EEG subcarriers was computed in this research, and active units (AU) were identified from respondent facial recordings. After that, a number of characteristics were implemented. The classifiers were hidden Markov model (HMM) and GentleBoost. The integration of facial recordings and EEG data increased the accuracy, according to the findings.

9.3 METHODOLOGY

9.3.1 Design

The suggested paradigm in Figure 9.4 depicts the overall flow of our research. The EEG signals were initially gathered, then data was preprocessed for this study. The key predictors were then used to identify bands of particular frequencies. Following that, appropriate characteristics were collected and chosen to be input into the classifier. Finally, the chosen characteristics were classified using SVM.

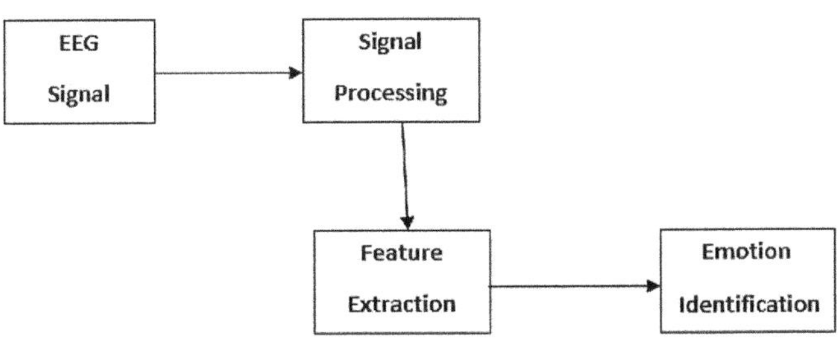

FIGURE 9.4 Proposed methodology flowchart.

9.3.2 Data Acquisition

The DEAP repository, which is a collection of data for emotion recognition utilizing biomedical parameters categorized utilizing a valence-arousal-dominance emotion paradigm, was utilized in this research [20]. There are 32 people in the DEAP dataset. One-minute videos with music were broadcast for every individual to activate the visual and auditory brain. Every person was given 40 videos, and seven distinct modalities was captured; EEG was utilized in this research, and additional details may be found on their website. The duration of the 40 video segments was chosen so they would fit within the purview. Every respondent was asked to rate each video clip on valence, arousal, dominance, and liking on a scale of 1 to 9. The arousal or valence label is significant if the score is higher than 4.5 and low if the score is much less: i.e., below 4.5. All of the impulses was captured at a sample rate of 512 Hz.

9.3.3 Data Analysis

- **Signal Processing**

The input is resampled to a sampling rate of 128 Hz. The computation of feature parameters takes less time with a reduced sampling frequency. Electrooculography (EOG) is no longer used since eye movement is the main source of noise. To eliminate sounds below 4 Hz or greater than 45 Hz, a bandpass filter is used. The average mean reference (AMR) technique is used to decrease electrical amplification, power line, and external interference noise. The average is computed for every chosen channel and deducted from each observation within the channel. A 60-second time frame is created from the noise-free EEG data. All sections go through the extracting features procedure, with 30% going into the classifier's knowledge base and 70% going into the classification task.

Earlier research has shown that data about feelings is primarily stored in the frontal and temporal regions of the brain. Furthermore, in order to reduce the computing expenses for the suggested approach, researchers mainly dealt with $Fp1$, $F3$, $F7$, $FC5$, $FC1$, $Fp2$, Fz, $F4$, $F8$. $FC6$, and $FC2$ channels that are linked to the frontal lobe of the brain. Classifiers and features from an array were time consuming due to the difficulty in manipulating the data to meet our needs. As a result, the preprocessed data was divided into 40 records, each of which represented a different video clip from the DEAP data collection. Every video included an array with a size of 8,064 x 352, with rows representing data length and columns representing the total number of channels for the 32 respondents.

- **Feature Extraction**

Because of discrete wavelet transform's (DWT) excellent multi-resolution capabilities in the analysis of nonstationary signals, it was used to extract features from windowed EEG signals of the chosen channels. The EEG data are windowed to increase the chances of detecting emotional states quickly. Theta (4 - 8 Hz), alpha (8 - 16 Hz), beta (16 - 32 Hz), gamma (32 - 64 Hz), and noises (> 64 Hz) are dissected from EEG data using the db4 function, also referred to as a mother wavelet function. Following

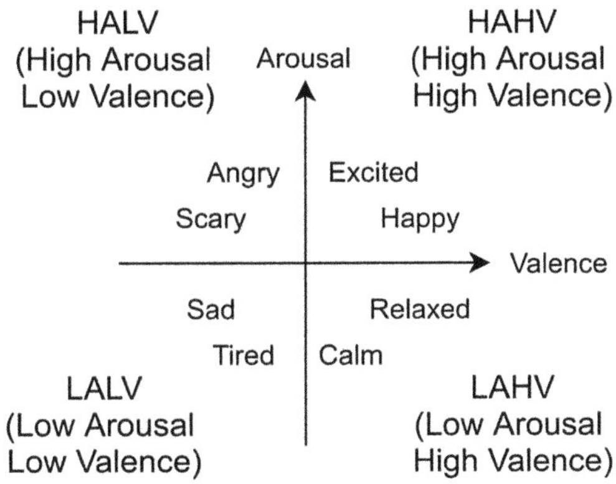

FIGURE 9.5 The four quadrants of the arousal-valence model [21].

that, each window of each frequency band's entropy and energy were retrieved. Each movie was assessed to fit into one of the four emotional quadrants as shown in Figure 9.5.

After sorting the films by quadrant and aggregating the groups of almost all the movies out of every quadrant, four video files containing the average values of the obtained bands were produced. The band frequencies then were adjusted so huge band frequencies had little effect on the classification model. The characteristics of the input signals were recovered after the band frequencies had been transformed. Based on the location or central tendency, the dispersion or spread, and the shape of the distribution, we derived the probabilistic characteristics' lower limit, upper limit, variability, variance, entropy of wave, and energy bandwidth for our research.

- **Emotion Classification**

There has been much research on machine learning methods, with SVM being one of the most effective classifiers for categorizing feelings. The SVM's fundamental notion is always to find a deciding hyperplane in order to divide information or data into two groups. The distances between the closest data points of both the categories as well as the hyperplane were maximized to find the best hyperplane for distinguishing two classes. The classification procedure uses the k-fold cross-validation method, which forecasts a confusion matrix model by separating the data samples into training and test datasets for training and testing, respectively. This technique divides the dataset into k equal sections at random and then does it again and again. Each time, a subset of size k is used as the test set, and a training set is created by combining the remaining k-1 subsets.

9.4 RESULTS AND DISCUSSION

In order to construct the confusion matrix model in our work, possible kinds of characteristics have been utilized for training and testing the classifier model. The performance was then determined using the model and k-fold cross-validation. The

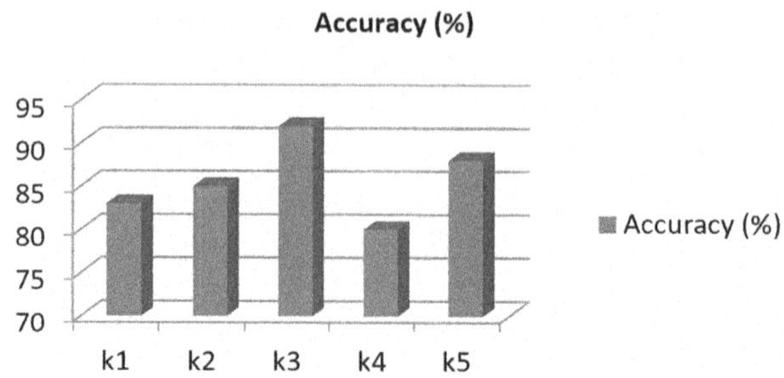

FIGURE 9.6 Graphical representation of accuracy considering different kernels.

grid-search technique was used to choose the features, kernel, and regularization that was then used to combine SVM using ten-fold cross-fold evaluation. The “libsvm” library is being used to design SVM, which would be a commonly used framework for the support vector machine. It was not simple to classify the statistical characteristics in order to get a good result. Because the first experiments did not provide satisfactory results, many factors had to be examined before coming to a decision. In this work, ten-fold cross-validation has been combined with the classification model having SVM classifier utilizing a method to determine the regularization and kernel parameters. The method was utilized to evaluate the accuracy of the classification method using an algorithm known as k-fold classification. The equation for accuracy is defined as shown. (Also see Figure 9.6.)

$$Accuracy = \frac{TP + TN}{TP + TN + FP + FN} \times 100$$

Where,

TP : the number of true positive

TN : the number of true negative

FP : the number of false positive

FN : the number of false positive

The dataset in consideration demonstrates the variance in classification accuracy for different kernels of the SVM classifier. When compared to other kernels, k4 classifiers have the lowest classification accuracy. The kernel k3 achieves the highest classification accuracy of 92%, as shown in Figure 9.6.

9.5 CONCLUSION

The study describes an original concept for an EEG-based emotion categorization system. Three factors – comparative waveform energy, comparative waveform entropy, and a new parameter (i.e., combination of standard deviations and discrete wavelet transform coefficients) – are computed employing the DEAP dataset as input data. An SVM classification method is also created with the goal of improving prediction performance. As a consequence, the package combining all three characteristics yields the greatest accuracy. It shows that perhaps the classification can continuously improve the emotion recognition system.

Different people may be thinking about different topics. Even when individuals are prompted to feel a certain emotion, other ideas and cognitive processes may still have a significant impact on their brain activity. Music or video data may be used to elicit emotions in a person's brain. Other methods may be employed to improve noise or artifact reduction performance. In the future, better feature extraction and categorization methods may be used. The work can be further enhanced to include features such as detection of mental state and controlling devices such as automatic music players based on the user's emotions, amongst many others. The classification method could be adjusted to return a greater precision using the custom tailor kernel function in the coming years. Other algorithmic permutations may also be tried.

REFERENCES

[1] M. Zhao, H. Gao, W. Wang, and J. Qu, "Research on human-computer interaction intention recognition based on EEG and eye movement," *IEEE Access*, 2020, doi: 10.1109/ACCESS.2020.3011740.

[2] D. Singh and M. Kumar, "New age vision detection technology," *J. Crit. Rev.*, vol. 7, no. 1, pp. 662–667, 2020.

[3] M. S. Solanki, "Sentiment analysis of text using rule based and natural language toolkit," *Int. J. Innov. Technol. Explor. Eng.*, vol. 8, no. 12S, pp. 164–168, 2019, doi: 10.35940/ijitee.l1049.10812s19.

[4] C. Li, "Deep reinforcement learning," *Reinf. Learn. Cyber Phys. Syst.*, vol. 11, no. 10, pp. 125–154, 2019, doi: 10.1201/9781351006620-6.

[5] Landauskas, Mantas, and Ugnė Orinaitė. "Novel Feature Extraction Technique Based on Ranks of Hankel Matrices with Application for ECG Analysis." *Mathematical Models in Engineering* 7, no. 2 (June 30, 2021): 40–49. https://doi.org/10.21595/mme.2021.22138.

[6] R. Murugeswari, C. Ramachandran, R. K. Dhanaraj, S. Yadav, M. R. Sundarakumar, and S. Basheer, A study on big data and IoT in diabetic retinopathy diagnosis and prediction. *Adv. Sci. Technol. Secur. Appl.*, vol. 3, pp. 305–315, 2018.

[7] M. Paliwal, "Brain tumor detection by fusing machine learning and neural network practices," *Int. J. Innov. Technol. Explor. Eng.*, vol. 8, no. 12S, pp. 108–111, 2019, doi: 10.35940/ijitee.l1032.10812s19.

[8] N. S. Zulpe and V. P. Pawar, "Review on brain tumor segmentation and classification techniques," *Int. J. Eng. Res.*, vol. 6, no. 11, pp. 1741–1746, 2017, doi: 10.17577/ijertv6is110008.

[9] S. Tuckett, "Triune brain," 2019. https://sarahtuckett.com.au/why-i-get-you-to-move-and-breathe-in-your-session/triune-brain/ (accessed Jul. 12, 2021).

[10] Guest6e2d539, "Limbic system emotional brain." www.slideshare.net/guest6e2d539/bizu10989/27-Limbic_System_Emotional_Brain (accessed Jul. 12, 2021).

[11] Mayoclinic, "EEG (electroencephalogram)." www.mayoclinic.org/tests-procedures/eeg/about/pac-20393875 (accessed Oct. 22, 2022).

[12] A. Weinreich, T. Stephani, and T. Schubert, "Emotion effects within frontal alpha oscillation in a picture oddball paradigm," *Int. J. Psychophysiol.*, 2016, doi: 10.1016/j.ijpsycho.2016.07.517.

[13] S. Nasehi and H. Pourghassem, "An optimal EEG-based emotion recognition algorithm using Gabor features," *WSEAS Trans. Signal Process.*, vol. 8, no. 3, pp. 87–99, 2012.
[14] K. Ishino and M. Hagiwara, "A feeling estimation system using a simple electroencephalograph," 2003, doi: 10.1109/icsmc.2003.1245645.
[15] A. R. Hidalgo-Muñoz et al., "Individual EEG differences in affective valence processing in women with low and high neuroticism," *Clin. Neurophysiol.*, 2013, doi: 10.1016/j.clinph.2013.03.026.
[16] S. Issa, Q. Peng, X. You, and W. A. Shah, "Emotion assessment based on EEG brain signals," in Abraham, A., Cherukuri, A., Melin, P., and Gandhi, N. (eds) *Intelligent Systems Design and Applications. ISDA 2018 2018. Advances in Intelligent Systems and Computing, vol 941*. Springer, Cham. doi: 10.1007/978-3-030-16660-1_28.
[17] S. H. Adil, M. Ebrahim, K. Raza, and S. S. Azhar Ali, "Prediction of eye state using KNN algorithm," 2018, doi: 10.1109/ICIAS.2018.8540596.
[18] X. Jie, R. Cao, and L. Li, "Emotion recognition based on the sample entropy of EEG," 2014, doi: 10.3233/BME-130919.
[19] S. Koelstra and I. Patras, "Fusion of facial expressions and EEG for implicit affective tagging," *Image Vis. Comput.*, 2013, doi: 10.1016/j.imavis.2012.10.002.
[20] S. Koelstra, C. Muehl, M. Soleymani, J.-S. Lee, A. Yazdani, T. Ebrahimi, T. Pun, A. Nijholt, and I. Patras, "DEAP: A database for emotion analysis; using physiological signals," *IEEE Trans. Affect. Comput.*, vol. 3, no. 1, pp. 18—31, 2012. doi: 10.1109/T-AFFC.2011.15.
[21] K. Suzuki, T. Laohakangvalvit, R. Matsubara, and M. Sugaya, "Constructing an emotion estimation model based on EEG/HRV indexes using feature extraction and feature selection algorithms," *Sensors*, 2021, doi: 10.3390/s21092910.

10 Prediction of Neonatal Mortality from Jaundice Using Machine Learning

Mayank Srivastava, Yajur, Sujata

10.1 INTRODUCTION

Infants born prematurely often develop jaundice because their livers are unable to process bilirubin in a way that allows it to be excreted in the urine. Bilirubin is a byproduct of hemolysis, the destruction of red blood cells. The skin and whites of the eyes of a baby with neonatal jaundice will be yellow [1]. Potential side effects include convulsions, CP, and kernicterus. When a newborn's bilirubin level is more than 308 mol/L (18 mg/dL) for more than two weeks, the condition is considered pathogenic jaundice. Two key factors [2] contribute to infant jaundice. First, since the liver's metabolic pathways are not fully developed until adulthood, the conjugation and removal of bilirubin are delayed during the shift from foetal to adult haemoglobin. Jaundice is caused by an excess of bilirubin in the blood (hyperbilirubinemia). Second, if phototherapy (see Figure 10.1) fails to alleviate the baby's jaundice, doctors should consider paediatric liver problems. There is a lack of bile ducts in diseases including Alagille syndrome, alpha 1-antitrypsin deficiency, and progressive familial intrahepatic cholestasis. An infant with persistent jaundice should see a doctor right away [3].

Jaundice is caused by hyperbilirubinemia, which affects around 60% of full-term newborns and almost 100% of preterm neonates [4]. Infant jaundice is usually nonthreatening and may be safely ignored. Early and precise diagnosis lowers an infant's chance of developing kernicterus [5]. When bilirubin builds up in the brain, a disease known as kernicterus may be developed [6]. Babies at risk of developing severe hyperbilirubinemia or kernicterus must be identified as soon as possible so that treatment can begin. Because an infant's central nervous system is still developing, protecting it from dangerously high bilirubin levels has become a top priority for paediatricians [7]. Specialized monograms are used to estimate the possibility of newborn jaundice by taking into consideration the baby's age, the blood or transcutaneous bilirubin levels, and other risk markers [8]. Several studies have demonstrated that even with improvements in risk assessment methods, bilirubin encephalopathy and kernicterus still occur in neonates [9].

Accurate diagnosis and the avoidance of its consequences are made possible by understanding the severity of jaundice in newborns and the modifiable risk factors associated with it. More and more newborns are being sent home from the hospital within the first 48 hours of life [10], so it's possible that jaundice won't have had a

DOI: 10.1201/9781003477280-10

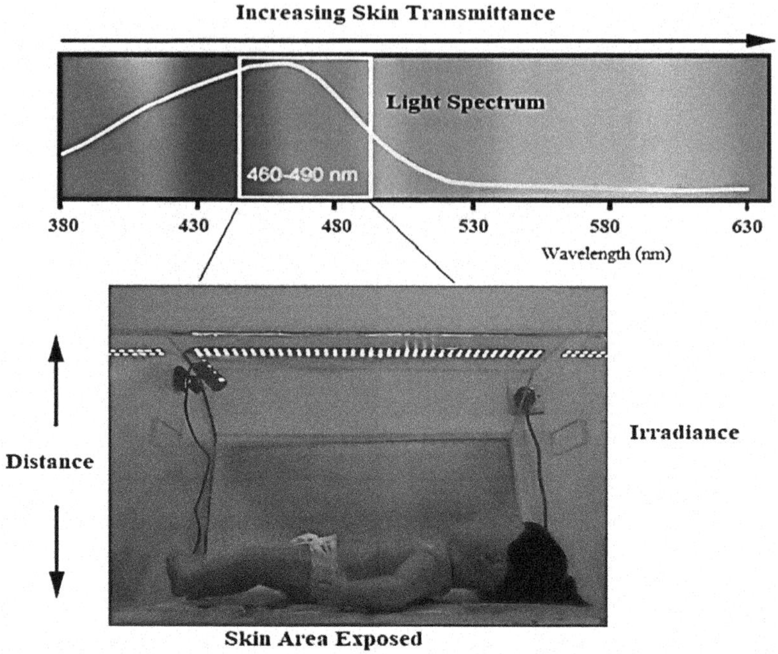

FIGURE 10.1 Neonatal jaundice detection.

chance to set in before Mom and Dad leave. Approximately 700 newborns per day, or 48 per every 1,000 live births, die in Nigerian hospitals due to complications from kernicterus. Children under the age of five in Nigeria have a high risk of dying due to premature birth [11]. Hyperbilirubinemia in infants may be prevented with early therapy [12, 13]. Data mining is a new area of study in computer science that employs several statistical techniques, database management systems, artificial intelligence (AI), and pattern recognition technologies (a subfield of machine learning).

Data mining has the ability to reveal previously unknown details about some illnesses and provide leads for further investigation in a variety of medical fields. The extraordinary accuracy of the generated models [14, 15] is an illustration of the value of data mining in the healthcare industry. Data mining has been shown to advance past knowledge and enhance the outcomes of other approaches [16], and it has been widely used in many fields of medicine. In recent decades, machine learning techniques have become more significant with developments in computational control in numerous domains, such as big data analytics and medical knowledge discovery [17]. In machine learning, deep learning refers to a collection of techniques for building deep hierarchical representations of data quickly [18]. These representations may be used to perform classification tasks more quickly. Industries that deal with huge amounts of ordinal data have found extensive applications for deep learning [19].

Large-scale medical image collections must be gathered and analyzed by clinicians using modalities like X-ray, CT, and MRI in order to provide high-quality diagnostic imaging services. Deep learning algorithms employed with big data have recently concentrated on the large amounts of patient data produced by healthcare systems. In this work, we utilize deep learning methods to classify neonatal jaundice severity by taking into account data related to avoidable risk factors. Predicting when a baby may develop jaundice using machine learning is the primary goal of this research [20].

10.2 LITERATURE REVIEW

Munkholm et al. [21] suggested use of iPhone 6's camera as the main data collector. The built-in dermatoscope of the iPhone 6 allows users to study skin lesions without being impeded by reflections from the skin's surface thanks to its magnification, powerful light, and clear viewing plate. In light of these findings, it was proposed that pictures of the glabella taken with a smartphone dermatoscope be utilized as a screening tool for hyperbilirubinemia in newborns. However, its usefulness was restricted [22]. Scientists have shown that noninvasive image processing algorithms may detect jaundice in babies. Face area, eye, and mouth identification led to the creation of the normalized RGB colour and threshold value.

Endang et al. [23] suggested a procedure for determining the probable danger zone. Babies' photos are taken using digital cameras, refined with a median filter to reduce grain, and colour corrected with a red-green-blue (RGB) colour card. The suggested approach only had a 67% success rate, but it was simple to execute and had a high correlation coefficient. The technique for measuring bilirubin levels in neonates was developed by Aune A. et al. [24]. The concept blends a physical model of how light flows to the skin with digital image colour analysis. The researchers created a sizable library of simulated reflectance spectra of newborn skin using a mathematical model based on diffusion theory. Additionally, the writers used a handheld digital camera that only had red, green, and blue color capabilities.

Rong et al. [25] integrated a colour calibration card with an automated smartphone. Taylor et al. [26] proposed another research that used the iPhone 5s to take photographs of 530 infants of African American, Hispanic American, and Asian American origin [27, 28]. The findings demonstrated that BiliCam may be used as a screening tool to assist evaluate whether neonates need blood collection for TSB, even though it is not accurate enough to be employed as a stand-alone technique for detecting jaundice in infants. The difficulty with the suggested method is that it requires a server and internet connectivity to operate [29–31].

10.3 MATERIALS AND METHODS

In this chapter, we classify the potentially fatal conditions of septic shock, birth asphyxia, necrotizing enter colitis (NEC), and respiratory distress syndrome using various machine-learning approaches. Figure 10.2 displays the complete approach.

The gathering and examination of pertinent data has commenced. Purifying data, dealing with missing values, and transforming data are all examples of preprocessing

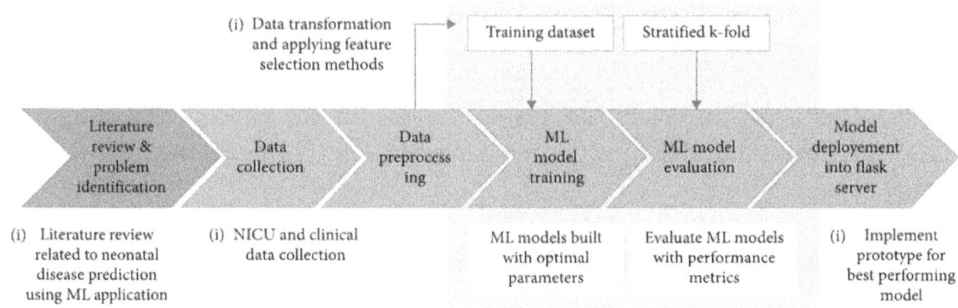

FIGURE 10.2 Workflow of the machine learning approach.

tasks performed in the proposed work. In order to find useful features, recursive feature reduction with a cross-validation approach was used. The cleaned data was sent to SVM, RF, and XGB, where it was further processed. The combined findings are the product of three separate models. The next section will go into further detail on these methods and approaches.

10.3.1 Modeling

When used properly, multilayer stacking–based machine learning methods outperform in terms of accuracy. The result of one model is fed into a third, more complex model (such as a meta-learner) in the stack [32]. Multiple classifiers or models (M1, M2, . . . , Mn) are used on the same dataset (S) [33]. Where xi is a pair of feature vectors and yi is a classification label, S is the set of all possible combinations of the two (yi). Research begins with building a collection of classifiers, C1, . . . , Cn, using the formula Ci = Mi (s). Each lower-level classifier's output is piped up to the meta-level learner. The training set for the meta-level classifier was constructed with the help of cross-validation.

Authors successfully stacked with and without feature selection by integrating support vector machines, random forests, and XGBoost, three fundamental learners. Figure 10.3 depicts this model-building procedure, and the parts that follow are prepared with newcomers in mind.

In Figure 10.3, classification and regression accuracy are equally well served by the SVM family of learning algorithms. Perhaps it contains several nonlinear structures. Finding a fair way to divide the economic groups is essential. As can be seen in Figure 10.4, SVM performs the classification process by constructing a hyperplane or a group of hyperplanes in a high-dimensional space. Information outside the hypercube's boundaries is represented by the extra data points [34]. To improve classification accuracy, move the hyperplane farther away from the nearest training data points. This is why a larger margin of error results in more precise data classification. One-vs-rest (OVR) and support vector machine (SVM) classifiers were employed in this research. The synergy between OVR and SVM made it attractive to combine the two for use in multiclass classifications [35].

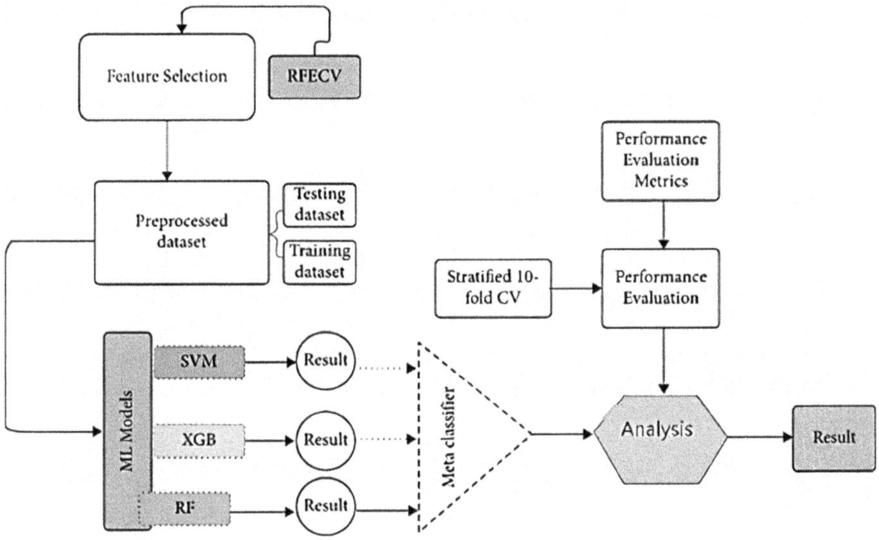

FIGURE 10.3 Flowchart of model building.

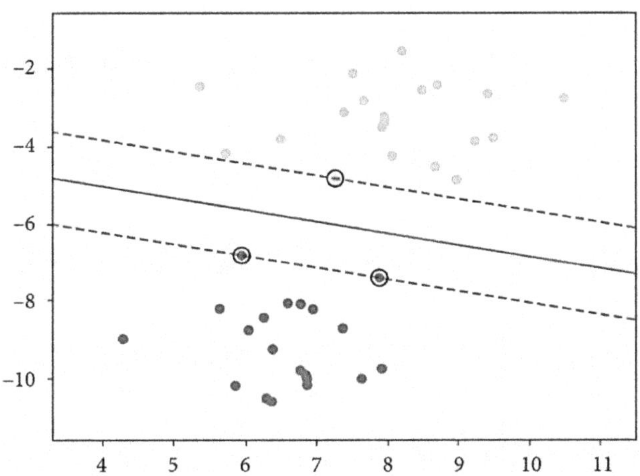

FIGURE 10.4 The linearly separable problem's decision function.

In Figure 10.4, classification and regression issues may be addressed with the aid of a decision tree and an ensemble of classifiers. A random decision tree forest is created using this method. The XGBoost project aimed to solve this problem by developing and implementing extended gradient-boosting decision trees that are acquiescent to machine learning. Classification and regression issues are amenable to XGBoost's solution.

10.3.2 Hyperparameter Tuning

Hyperparameter tuning refers to the process of selecting a collection of hyperparameters with the intention of improving performance. Modifications may be made manually or automatically. The current study benefited from the use of a robotic technique. Grid search and randomized search are the most often used algorithms. Since the grid search performs a global search over a fixed section of the algorithm's hyperparameter space, it is a popular method for hyperparameter optimization. In the training phase, candidates are assembled from scratch using inputs you provide. Problems arise when trying to use this strategy in environments with more than three dimensions. When the performance of a machine learning system is dependent on a small set of tuning elements, grid searches perform poorly in comparison to random searches. For this reason, we used a random search strategy.

10.4 RESULTS AND DISCUSSION

The findings of the stacking model were compared to those of other models. A flask server has been used to deliver the most efficient version of the model.

The dataset comprises 20 characteristics, the largest of which is the target class, and a total size of 2,298. The study counts four common infant diseases – septicemia, RDS, NEC, and paternal asphyxia – among the most common (PA). Additionally, there were 412 instances of necrotizing enterocolitis, 527 cases of maternal asphyxia, 648 cases of respiratory distress syndrome, and 711 cases of sepsis. Fifty-nine percent of pregnant women continue to get prenatal care. According to the data, only 49% of infants were born on time, while 4.6% were born prematurely, and 46.1% were delivered late. "Antenatal follow-up" refers to a pregnant woman's care and monitoring by medical staff.

Jaundice is a medical disorder characterized by a yellowing of the skin and the whites of the eyes. An excess of bilirubin, a yellow pigment produced when red blood cells are broken down, is the underlying cause of this condition. Factors including age, sex, and general health may influence how the body distributes red blood cells and white blood cells. In comparison to the approximately four to six million red blood cells found in a microliter of adult blood, there are approximately 4,000 to 11,000 white blood cells. The screening laboratory's reference ranges may cause some adjustments to these numbers.

The HADM ID is the patient's unique identifier in the MIMIC-III database, which is often utilized in intensive care studies. The total number of patients and the total number of hospitalizations will affect the visual look of the distribution of HADM IDs in a dataset.

The ratio of albumin to globulin in the blood is measured by a lab test abbreviated as ALB/GLB. Jaundice is a common reason for requesting this test, which is used to assess liver health. The number of patients included and the results of relevant tests determine the ALB-to-GLB ratio for a specific dataset. In most cases, a value between 1.0 and 2.5 for the ALB-to-GLB ratio is desirable. Liver disease and other abnormalities may be indicated by abnormal readings.

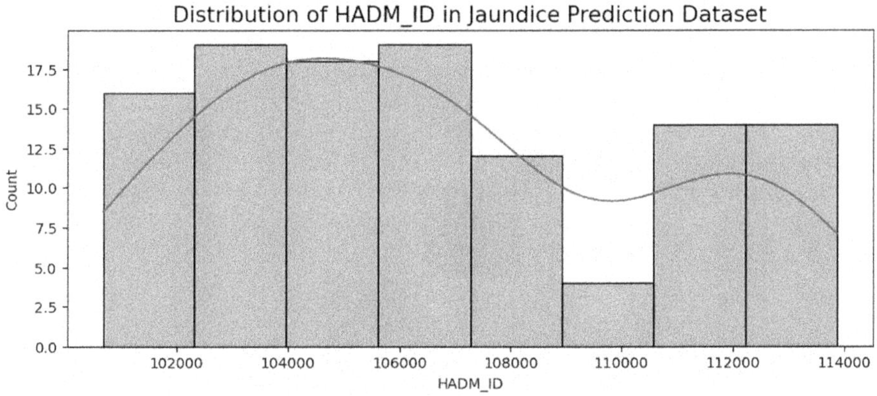

FIGURE 10.5 Distribution of HADM_ID in jaundice prediction dataset.

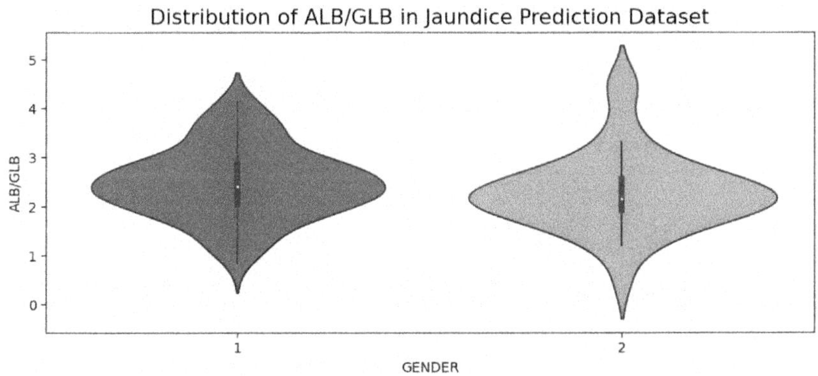

FIGURE 10.6 Distribution of ALB/GLB in jaundice prediction dataset.

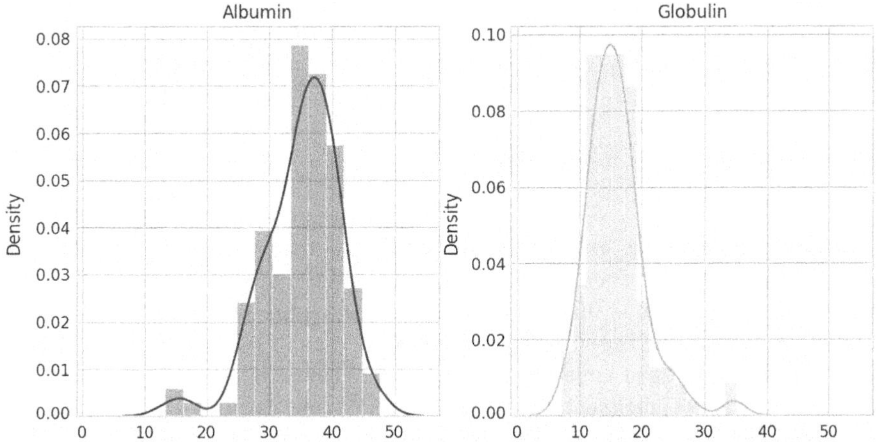

FIGURE 10.7 Albumin and globulins.

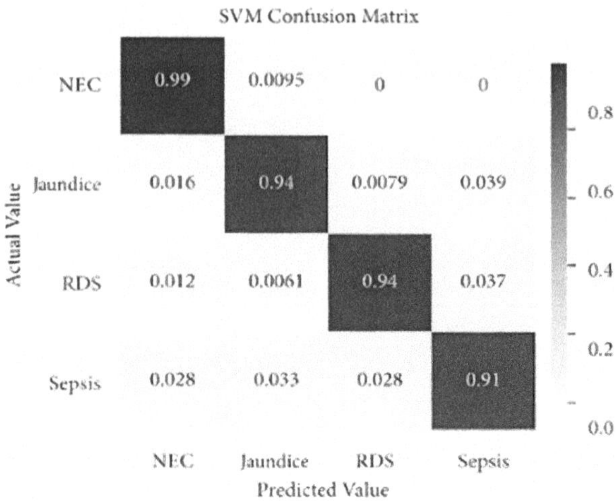

FIGURE 10.8 Confusion matrix for SVM.

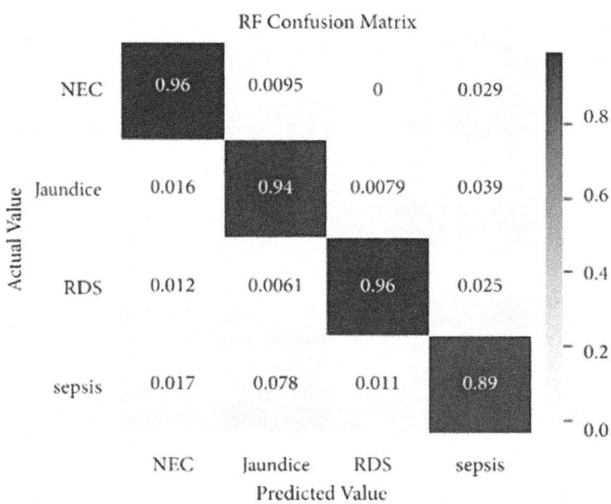

FIGURE 10.9 Confusion matrix for RF.

To train the model, recursive feature removal and cross-validation (RFECV) were utilized. That's why current research narrowed down the focus to just these 12 essential factors. Figure 10.8 is the confusion matrix showing how precise the SVM is. Correct diagnoses have been made in 104 of 105 cases of NEC, 154 of 163 cases of RDS, 119 of 127 cases of jaundice, and 164 of 180 cases of sepsis. Five instances of PA were incorrectly labelled as NEC, one as RDS, and another as PA. Similarities also exist between the other misclassifications shown in the illustration.

The effectiveness of random forest may be measured by a confusion matrix, like the one shown in Figure 10.9. In a study of 105 patients with NEC, 92% were correctly diagnosed; in a study of 120 patients with PA, 100% were correctly diagnosed;

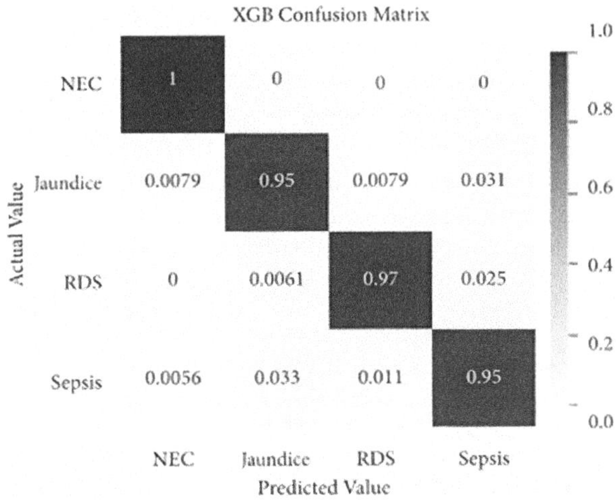

FIGURE 10.10 Confusion matrix for XGB.

TABLE 10.1
Comparisons of Performance with SVM, RF, and XGB

Models	Precisions	Recall	F1-Score	Accuracy
SVM	95.1%	95.7%	95.1%	94.9%
RF	96.8%	96.7%	95.7%	95.8%
XGB	96%	96.1%	95.8%	95.9%

in a study of 158 patients with RDS, 100% were correctly diagnosed; and in a study of 180 patients with sepsis, 163 were correctly diagnosed. Two patients with NEC were incorrectly identified with sepsis by RF while only one was incorrectly labelled with PA. The image shows this, along with the other incorrect labels. The normalized confusion matrix displays the proportion of correct identifications as a decimal.

Figure 10.10 showcases the XGB classifier's output. There are a total of 105 instances of Sepsis, 121 cases of NEC, 154 cases of RDS, and 163 cases of PA that were diagnosed accurately. It's possible the sample uses inaccurate classifications.

Using recursive feature reduction and cross-validation, accuracy was improved in all three cases. Positive identification was obtained in the following instances: overall, sepsis accounted for 171 of the 180 cases, whereas NEC caused 105, jaundice caused 123, RDS caused 158, and sepsis caused 105. Incorrectly labelled data seldom occurs. Table 10.1 displays the results of tests that compare SVM, RF, XGB with recursive feature removal, and a ten-fold cross-validation.

10.5 CONCLUSION

One significant finding suggests that training ML models with a larger number of base models may improve their effectiveness. Several studies have looked at the best ways to improve forecast precision. Predicting neonatal disease requires a number of indicators, including but not limited to C-reactive protein, Apgar score, resuscitation, whiteout, low lung volume, intercostal subcostal retractions, oxygen saturation, blood cultures, gestational age, seizures, respiratory rate, weight, white blood cell count, grunting, and age at birth (jaundice). When compared to the other simple reference models, the ML approach performs better regardless of whether or not feature selection is used (random forest, support vector machine, and XGB). The detailed analysis achieves a 95.79 f1 score, 96.13% recall, 96.01% precision, and 95.91% accuracy.

REFERENCES

[1] Dzulkifli, F. A., Mashor, M. Y., Khalid, K. (2018). Techniques for deciding bilirubin level in neonatal jaundice screening and observing: A writing survey. *J. Eng. Res. Educ.* 10, 1–10.
[2] Brits, H., Adendorff, J., Huisamen, D., Beukes, D., Botha, K., Herbst, H., Joubert, G. (2018). The commonness of neonatal jaundice and hazard factors in solid term children at Public Region Clinic in Bloemfontein. *Afr. J. Demure. Med. Care Fam. Drug.* 10, 1–6.
[3] Bhutani, V. K., Zipursky, A., Blencowe, H., Khanna, R., Sgro, M., Ebbesen, F., Chime, J., Mori, R., Slusher, T. M., Fahmy, N. (2013). Neonatal hyperbilirubinemia and Rhesus infection of the infant: Frequency and hindrance gauges for 2010 at local and worldwide levels. *Pediatr. Res.* 74, 86–100.
[4] Vodret, S. (2016). *Neonatal Hyperbilirubinemia: In Vivo Portrayal of Components of Bilirubin Neurotoxicity and Pharmacological Medicines.* New Delhi, India: Worldwide Place for Hereditary Designing and Biotechnology.
[5] Mishra, S., Agarwal, R., Deorari, A. K., Paul, V. K. (2008). Jaundice in the babies. *Indian J. Pediatr.* 75, 157–163.
[6] Hyperbilirubinemia, A. A. o. P. S. o. (2004). The board of hyperbilirubinemia in the baby at least 35 weeks of growth. *Pediatrics* 114, 297–316.
[7] Mantagou, L., Fouzas, S., Skylogianni, E., Giannakopoulos, I., Karatza, A., Varvarigou, A. (2012). Patterns of transcutaneous bilirubin in children who foster huge hyperbilirubinemia. *Pediatrics* 130, e898–e904.
[8] Mansouri, M., Mahmoodnejad, A., Taghizadeh Sarvestani, R., Gharibi, F. (2015). An examination between transcutaneous bilirubin (TcB) and all out serum bilirubin (TSB) estimations in term children. *Int. J. Pediatr.* 3, 633–641.
[9] Alsaedi, S. A. (2018). Transcutaneous bilirubin estimations can be utilized to quantify bilirubin levels during phototherapy. *Int. J. Pediatr.* 2018, 4856390.
[10] Mreihil, K., Nakstad, B., Stensvold, H. J., Benth, J. Š., Hansen, T. W. R., Gathering, N. N. P. S., Organization, N. N., Scheck, O., Nordin, S., Prytz, A. (2018). Uniform public rules don't forestall wide varieties in the clinical use of phototherapy for neonatal jaundice. *Acta Paediatr.* 107, 620–627.
[11] Maisels, M. J., McDonagh, A. F. (2008). Phototherapy for neonatal jaundice. *N. Engl. J. Drug.* 358, 920–928.
[12] Mreihil, K., Benth, J. Š., Stensvold, H. J., Nakstad, B., Hansen, T. W. R., Gathering, N. N. P. S., Organization, N. N., Scheck, O., Nordin, S., Prytz, A. (2018). Phototherapy is normally utilized for neonatal jaundice however more prominent control is expected to keep away from poisonousness in the most weak newborn children. *Acta Paediatr.* 107, 611–619.
[13] Owaymir, A. D. A., Aseeri, R. M. A., Albariqi, M. A. A., Alalyani, M. S., Almansaf, J. A. A., Albalwi, A. B. K., ALSalem, R. A., Asiri, K. J., Baeyti, N. Y. H., & Alrobaie, K. A. (2021). An Overview on Diagnosis and Management of Neonatal Jaundice. *Archives of Pharmacy Practice*, 12(2), 99–102. https://doi.org/10.51847/1TW12LWtPn.
[14] Donel, J. (2019). Bili cover phototherapy. *Int. J. Contemp. Pediatr.* 6, 2231–2234.

[15] Maisels, M., Watchko, J., Bhutani, V., Stevenson, D. (2012). A way to deal with the administration of hyperbilirubinemia in the preterm baby under 35 weeks of development. *J. Perinatol.* 32, 660–664.
[16] Leo, M., Farinella, G. M., Furnari, A., Medioni, G. (2022). Editorial: Machine vision for assistive technologies. *Front. Comput. Sci.* 4, 937433.doi: 10.3389/fcomp.2022.937433. Accessed August 26, 2024.
[17] Hashim, W., Al-Naji, A., Al-Rayahi, I. A., Oudah, M. (2021). PC vision for jaundice location in youngsters utilizing realistic UI. *IOP Conf. Ser. Mater. Sci. Eng.* 1105, 012076.
[18] Leartveravat, S. (2009). Transcutaneous bilirubin estimation in full term youngster by computerized camera. *Prescription J. Srisaket Surin Buriram Hosp.* 24, 105–118.
[19] Mansor, M., Yaacob, S., Hariharan, M., Basah, S., Jamil, S. A., Khidir, M. M., Rejab, M., Ibrahim, K. K., Jamil, A. A., Junoh, A. (2012). Jaundice in infant observing utilizing variety location technique. *Procedia Eng.* 29, 1631–1635.
[20] Mansor, M., Hariharan, M., Basah, S., Yaacob, S. (2013). New infant jaundice checking plan in light of blend of pre-handling and variety discovery technique. *Neurocomputing* 120, 258–261.
[21] Munkholm, S. B., Krøgholt, T., Ebbesen, F., Szecsi, P. B., Kristensen, S. R. (2018). The cell phone camera as a likely technique for transcutaneous bilirubin estimation. *PLoS ONE* 13, e0197938.
[22] Kawano, S., Zin, T. T., Kodama, Y. (2018). A concentrate on non-contact and painless neonatal jaundice location and bilirubin worth forecast. In *Procedures of the 2018 IEEE Seventh Worldwide Gathering on Purchaser Hardware (GCCE)*, Nara, Japan, 9–12 October 2018, pp. 401–402.
[23] Juliastuti, E., Nadhira, V., Satwika, Y. W., Aziz, N. A., Zahra, N. (2019). Risk zone assessment of infant jaundice in light of skin variety picture examination. In *Procedures of the 2019 Sixth Global Meeting on Instrumentation, Control and Computerization (ICA)*, Bandung, Indonesia, 31 July–2 August 2019, pp. 176–181.
[24] Aune, A., Vartdal, G., Bergseng, H., Randeberg, L. L., Darj, E. (2020). Bilirubin gauges from cell phone pictures of babies' skin associated profoundly to serum bilirubin levels. *Acta Paediatr.* 109, 2532–2538.
[25] Rong, Z., Luo, F., Mama, L., Chen, L., Wu, L., Liu, W., Du, L., Luo, X. (2016). Assessment of a programmed picture based evaluating procedure for neonatal hyperbilirubinemia. Zhonghua trama center Ke Za Zhi/Jawline. *J. Pediatr.* 54, 597–600.
[26] Taylor, J. A., Bold, J. W., de Greef, L., Goel, M., Patel, S., Chung, E. K., Koduri, A., McMahon, S., Dickerson, J., Simpson, E. A. (2017). Utilization of a cell phone application to evaluate neonatal jaundice. *Pediatrics* 140, e20170312.
[27] Aydın, M., Hardalaç, F., Ural, B., Karap, S. (2016). Neonatal jaundice location framework. *J. Prescription Syst.* 40, 166.
[28] Padidar, P., Shaker, M., Amoozgar, H., Khorraminejad-Shirazi, M., Hemmati, F., Najib, K. S., Pourarian, S. (2019). Location of neonatal jaundice by utilizing an android operating system-based cell phone application. *Iran. J. Pediatr.* 29, e84397.
[29] Angelico, R., Liccardo, D., Paoletti, M., Pietrobattista, A., Basso, M. S., Mosca, A., Safarikia, S. (2020) A clever cell phone application for baby stool variety acknowledgment: A simple and viable device to distinguish acholic stools in babies. *J. Drug Screen.* 26, 230–237.
[30] Miah, M. M. M., Tazim, R. J., Johora, F. T., Al Imran, M. I., Surma, S. S., Islam, F., Shabab, R., Shahnaz, C., Subhana, A. (2019). Non-invasive bilirubin level quantification and jaundice detection by sclera image processing. In *2019 IEEE Global Humanitarian Technology Conference (GHTC)*, Seattle, WA, USA, 2019, pp. 1–7, doi: 10.1109/GHTC46095.2019.9033059.

[31] Outlaw, I. d., Nixon, M., Odeyemi, O., Macdonald, L. W., Compliant, J., Leung, T. S. (2020). Cell phone evaluating for neonatal jaundice through encompassing deducted sclera chromaticity. *PLoS ONE* 15, e0216970.

[32] Viau Colindres, J., Rountree, C., Destarac, M. A., Cui, Y., Pérez Valdez, M., Herrera Castellanos, M., Mirabal, Y., Spiegel, G., Richards-Kortum, R., Oden, M. (2012). Imminent randomized controlled concentrate on looking for a minimal expense drove and ordinary phototherapy for treatment of neonatal hyperbilirubinemia. *J. Trop. Pediatr.* 58, 178–183.

[33] Kolkur, S., Kalbande, D., Shimpi, P., Bapat, C., Jatakia, J. (2017). Human skin location utilizing RGB, HSV and YCbCr variety models. arXiv, arXiv:1708.02694.

[34] Bangare, S. L., Dubal, A., Bangare, P. S., Patil, S. (2015). Evaluating Otsu's technique for picture thresholding. *Int. J. Appl. Eng. Res.* 10, 21777–21783.

[35] Chowdhary, A., Dutta, S., Ghosh, R. (2017). Neonatal jaundice location utilizing variety discovery technique. *Int. Adv. Res. J. Sci. Eng. Technol.* 4, 197–203.

11 Implementation of ML Techniques for Heart Prediction in Healthcare

Tanvi Rustagi, Meenu Vijarania

11.1 INTRODUCTION

Machine learning has seen a rise in popularity over the past decades. In order to estimate and govern the accuracy of a specified dataset, machine learning algorithms incorporate different categories of classifiers from supervised learning, unsupervised learning, and ensemble learning [1]. We can use that knowledge in our research. Cheap memory and powerful processing capacity contribute to raising interest [2]. NLP, structure assembly, computer visualization, healthcare, and education are just a few of the many application areas in which machine learning is crucial [3]. This study seeks to identify cardiac illness. By using a patient's past therapeutic record, our methodology can identify those who are most expected to be diagnosed with a cardiac illness. Preoperative diagnostics, including treadmill testing, angiography, and ECO, have improved the ability to diagnose cardiac problems [4]. Regardless, the aforementioned diagnostic procedure has drawbacks. The mortality rate for people with heart disease is reduced when the problem is discovered early [5]. In order to identify cardiac disease at an early stage, advancements in methodologies are now accessible [6].

Chest distress or heaviness, shoulder or arm pain, breathing difficulty, and perspiration are the distinctive indications and warning signals of a heart attack. Cardiovascular disease is brought on by arterial blockages, and a totally clogged coronary artery will result in a heart attack. The heart receives blood from arteries. If the arteries are blocked, it reduces blood flow to the heart and impairs cardiac function.

11.1.1 Categories of Heart Illness

Coronary artery disease: This disease is brought on by the deposit of plaque in the arteries that carry blood toward the heart [7]. The arteries are blocked, which reduces blood flow to the heart and impairs cardiac function.

Heart arrhythmia: A disorder in which the heart pumps too slowly, too quickly, or irregularly as a result of an electrical system malfunction. Exhaustion, unconsciousness, vertigo, and breathing irregularity are the most typical signs of a peculiar heart rhythm.

DOI: 10.1201/9781003477280-11

Congenital heart disease: A condition in which the heart's ability to drive enough blood to fulfill the body's oxygen demands is reduced, which results in heart failure at an early age [8].

Heart valve disease: Valves at each of the heart's four chamber exits keep the blood flowing through the heart in a single direction. The four heart valves make sure that blood does not leak backward and flows smoothly forward. Backward leakage might be caused by dysfunctional heart valves. The medical name for his is heart valve disease.

This chapter offers a heart disease detection road map. The investigation makes use of the heart disease dataset. To determine if patients are likely to have heart disease, we classify them using 14 medical characteristics. The research demonstrates how machine learning techniques like logistic regression (LR), K-nearest neighbors (KNN), and random forest (RF) contribute to heart disease prediction [9]. These methods can aid in the accurate early identification of heart disease. The basis for performing a comparative examination of various methods is also provided in this chapter. Finally, while creating a unified intelligent model, this research is essential in determining the best machine learning approach.

There are four sections in this chapter. A literature review in Section 11.2 analyzes the work completed thus far. The intended strategy is covered in Section 11.3. The results are described and evaluated against other machine learning models in Section 11.4, which also contains the interpretation drawn from the suggested framework.

11.2 RELATED WORK

The use of supervised machine learning algorithms to identify cardiovascular heart illness has had an extensive impact on this work. A summary of the previous work done so far is provided in this chapter. Using a variety of methods, an effective prediction of cardiovascular disease has been achieved. Exposure to heart disease has been the subject of related studies, which are presented in this section. These studies used ML techniques, including KNN, support vector machine (SVM), random forest (RF), DT, CNN, and logistic regression on a variety of datasets, and they included comparison parameters like accuracy, recall, f1 score, sensitivity, specificity, and precision.

The random forest algorithm, with a 90.16% percent accuracy score for heart disease prognosis, is the most effective. This research evaluates the accuracy of the DT, random forest, logistic regression, and Naive Bayes algorithms for forecasting heart attacks [10]. Logistic regression is applied to predict outcomes. The suggested model consistently provides superior performance with a lower level of uncertainty than ANFIS, LR, KNN, RF, J48, GB, and ANN. The suggested model exceeds the competition for F-measure analysis and accuracy on current models by 1.2782% and 1.4765%, respectively [11]. Many researchers have utilized the Cleveland heart disease data record, which contains 76 characteristics and 303 occurrences, with just 14 attributes being used because of missing information. The number of characteristics needed to produce an accurate model can be decreased by using feature selection methods, which investigate the relationship between various traits and their influence on model accuracy. According to various study articles in the field, KNN and neural networks are generally highly accurate for predicting heart disorders [12]. The techniques for backpropagation neural networks, fuzzy KNN, fuzzy Naive Bayes, K-means clustering, and logistic regression were used. The medical investigation of heart issues uses a ten-fold cross-validation technique. The best accuracy was shown by the backpropagation neural networks, which had 98.2%, 87.64% recall, and 89.65% precision [13]. The FS algorithm Relief, coupled with LR and evaluated using ten-fold cross-validation, produced the highest accuracy rate of 89% among the classifiers. When ANN was employed with Relief, SVM with the feature selection technique mRMR demonstrated the best performance (88%), and sensitivity was 100% [14]. The gram polynomial and PNN model for characterizing cardiac disorders used PCG signals. The suggested model yielded 94% accuracy, 93% sensitivity, and 91% specificity, respectively [15]. A collection of models was proposed using the WEKA tool and supervised learning methodologies. The classification algorithms DT, KNN, RF, and NB were used to forecast the chance of getting heart disease. The KNN method produced the best accuracy [16]. An accurate source of information may be found in electronic medical records; use their health data and machine learning techniques. The prognosis of heart failure patients' chances of survival is improved by our physicians [17]. Several machine learning approaches, including ANN, DT, RF, SVM, NB, and KNN, are essential for early heart disease prediction. The choice of features determines how effective these algorithms are. Better results can be obtained using techniques for feature selection such as the grid search and random search algorithms [18]. The BPA-NB scheme is used in the suggested model to create groups and forecast illnesses, coupled with clustering and probabilistic classification based on the Bayes theorem. It provides independent assumptions between the features. The proposed model utilized Hadoop-spark as a big data processing tool. The BPA-NB scheme on the UCI machine learning repository gives an accuracy of about 97.12% [19]. The proposed model used the DLMNN classifier to identify heart illnesses; the substitution cipher along with the SHA-512 are used for authentication. The greatest level of security (95.87%) is obtained with the PDH-AES approach used for secure data transfer and the shortest possible time for both encryption and decryption. Outstanding values for sensitivity (95.59), accuracy (96.8), specificity (91.30), and f-measure (98.25) are attained by the DLMNN prediction [20]. A comparison of LR, K-next neighbor, SVM, RF, DT, and deep learning classifiers is carried out by the suggested method. With 94.2% accuracy, 83.1% specificity, and 82.3% sensitivity, deep learning outperformed other classifiers in the evaluation that was centered on these three metrics. The isolation forest technique is used for outlier discovery in real-world data issues. Larger datasets significantly improve both deep learning and conventional machine learning techniques. A notable improvement was seen when ANN architecture correctness was compared to reports from other researchers [21]. The UCI repository dataset was utilized to compare the accuracy of machine learning algorithms; evaluation criteria, including RMSE; precision; recall; and MA and accuracy. SVM, logistic regression, and ANN all produced comparable predictions. On the other hand, random forest outperformed all other algorithms in accuracy and other assessment considerations with an accuracy of 95.60% [22].

11.3 PROPOSED METHODOLOGY

This technique is a practice that implies stages that translate available data into approved data arrangements for awareness. The submitted approach consists of phases, with the first step being the gathering of data. Preprocessing is the next stage, in which we analyze the data. The third stage is feature selection, in which significant values are extracted [23].

Contingent on the procedures used, data preprocessing deals with omitted values, data cleansing, and standardization [24]. Preprocessed data records are then categorized using a supervised classifier. The classifiers working in the intended model are KNN, logistic regression (LR), and random forest (RF) supervised classifier [25]. The intended model is then implemented, and using a variety of performance measures, we assess our model's recall, accuracy, and F1 score. The majority of the experimental work in this chapter was devoted to classification methods. The task was divided into two phases. Using the given dataset, phase one of machine learning entails training the machines. Testing is done in phase two. Python programming is employed to create and test various supervised machine learning procedures, as shown in Figure 11.1.

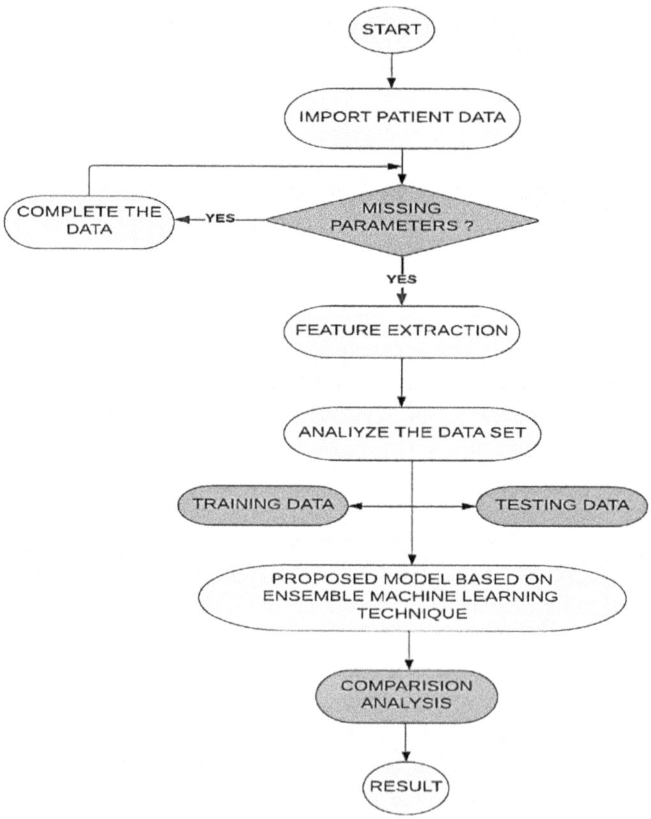

FIGURE 11.1 Workflow methodology

11.3.1 Data Gathering

Data gathering is also referred to as data collection. By concentrating on the history of cardiac issues and their relationship with other medical concerns, a structured dataset of individuals was chosen [26]. As a collective term, "heart diseases" refers to the various conditions that harm the heart. As reported by the World Health Organization, a large proportion of middle-aged adult deaths are caused by cardiovascular disorders [27]. We use a data record that contains the past medical examination data of 1,026 distinct patients, all of different age groups. The health constraints of the patient, such as age, blood pressure (BP), fasting sugar level (diabetic), etc. in this data record [28] allow us to determine if patients have been identified with heart illness or not. The heart disease data record was consumed for this experimental study. The number of items used from the data record was 1,026 [29], of which 526 were malignant (cardiac patients), and 500 were benign (noncardiac patients). The same is shown in Figure 11.2, where "0" represents benign, and "1" is malignant.

11.3.2 Data Preprocessing

Data preprocessing is done on a dataset to improve excellence by removing unnecessary data. Data cleansing, transformation, and reduction are the three processes in data preprocessing [30].

The dataset used for the analysis included 14 factors, including age, sex, cholesterol, fasting blood sugar, and many more. Identifying whether or not the data is balanced is crucial. It is evident in Figure 11.2 that the dataset is not uniformly balanced. The number of benign cells is nearly identical to the number of malignant cells [31]. Figure 11.3 demonstrates the relationship between all characteristics using a heat map.

11.3.3 Feature Selection

Additionally, by downsizing features from a higher dimension to a less dimensional region, feature selection approaches to aid in dimension reduction [32]. After the

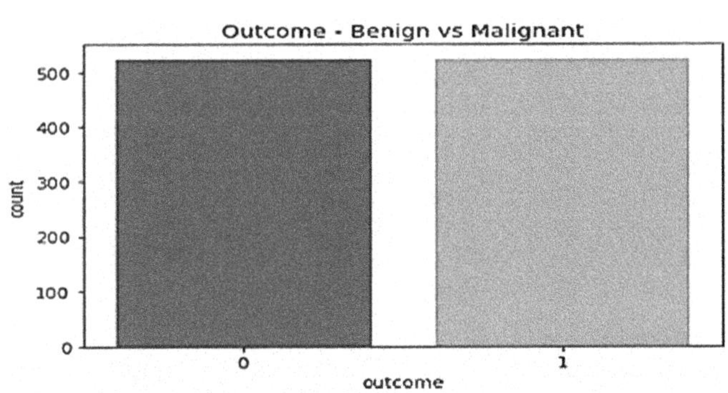

FIGURE 11.2 Representation of benign and malignant cells.

data selection procedure was complete, the chosen characteristics were extracted. Figure 11.4 shows the graph with the chosen appearances after the unnecessary characteristics have been disinterested [33].

11.3.4 Utilizing Machine Learning Models

Machine learning algorithmic models are used on the provided dataset in this phase. These algorithmic models will be educated and constructed in the present phase utilizing the processed data [34]. Random forest, logistic regression (LR), and KNN ML classification models have all been considered. The proposed model is tested to undertake a comparative study of performance.

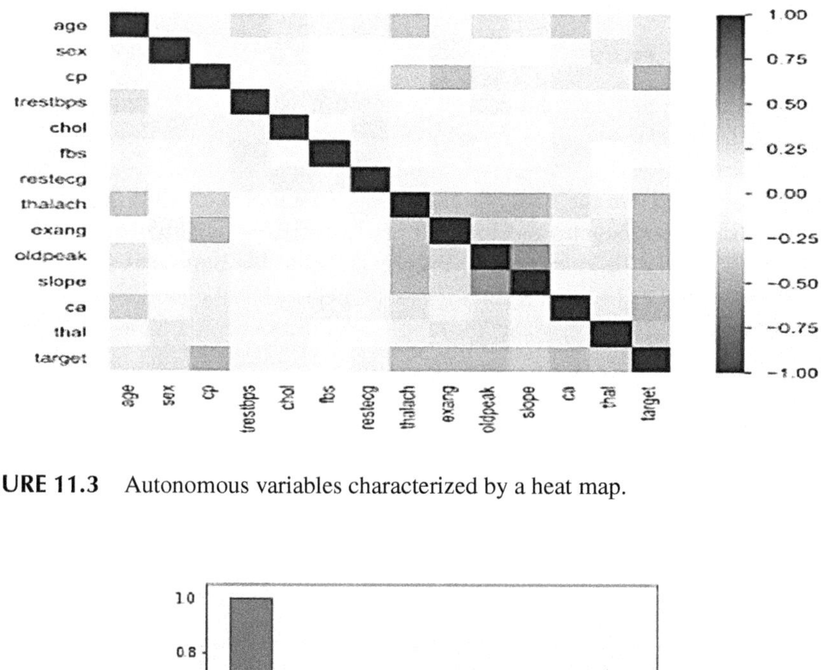

FIGURE 11.3 Autonomous variables characterized by a heat map.

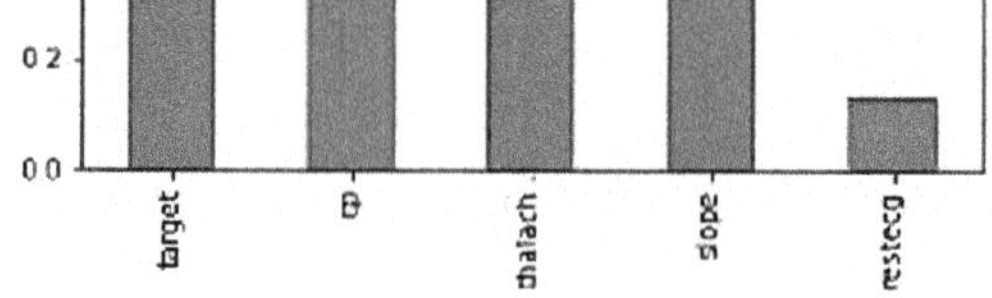

FIGURE 11.4 Dataset denotes specific attributes.

TABLE 11.1
Implementation Assessment of Different Machine Learning Algorithmic Models

Algorithm	Accuracy	Recall	F1 Score
K-NN	0.86	0.91	0.87
Logistic regression	0.87	0.93	0.88
Random forest	0.84	0.88	0.85

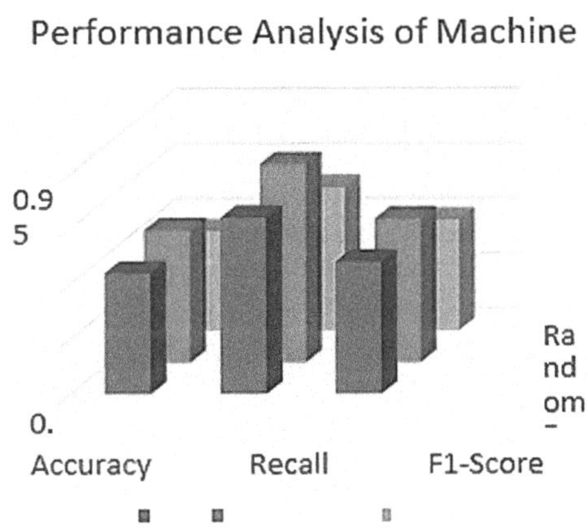

FIGURE 11.5 Performance analysis of machine learning model.

11.4 RESULT AND DISCUSSION

For the early diagnosis of heart illness, the intended model of machine learning algorithms is applied. In the proposed method, three supervised machine learning algorithms are taken into account. The Cleveland database heart disease dataset is used to implement logistic regression, random forest, and K-NN. Accuracy, recall, and F1 score have all been taken into consideration while evaluating the performance. (See Table 11.1.)

The results of various machine learning approaches are compared and evaluated against one another. The comparative analysis in Figure 11.5 was created using data from Table 11.1 and may aid in understanding. The assessment report states that logistic regression is excellent in accuracy (87%), recall (93%), and F1 score (88%).

11.5 CONCLUSION

Heart disorders are among the most common and deadliest disorders across the globe. A number of variables affect the likelihood of developing a disease, such as age, sex, cp, cholesterol, and fasting blood sugar. By analyzing patient medical histories, machine learning algorithms help spot diseases early. A supplementary dataset of 1,026 records and 14 attributes was used in this investigation. Supervised KNN classifier, supervised random forest classifier (RF), and supervised logistic regression (LR) classifier were the three algorithms used in the proposed model. With a maximum accuracy of 87%, recall of 93%, and F1 score of 88%, logistic regression outperformed the other methods in terms of accuracy. In the future, more research can be done on the range of available datasets, which can help doctors to identify the disease early.

REFERENCES

[1] A. Javeed, S. U. Khan, L. Ali, S. Ali, Y. Imrana, and A. Rahman, "Machine learning-based automated diagnostic systems developed for heart failure prediction using different types of data modalities: A systematic review and future directions," *Computational and Mathematical Methods in Medicine*, vol. 2022, pp. 1–30, Feb. 2022, doi: 10.1155/2022/9288452.

[2] M. J. N. Nayeem, S. Rana, and M. R. Islam, Prediction of heart disease using machine learning algorithms. *European Journal of Artificial Intelligence and Machine Learning*, vol. 1, no. 3, pp. 22–26, 2022. doi: 10.24018/ejai.2022.1.3.13.

[3] A. Pandita, S. Vashisht, A. Tyagi, and S. Yadav, "Prediction of heart disease using machine learning algorithms," *International Journal for Research in Applied Science and Engineering Technology (IJRASET)*, vol. 9, no. V, pp. 2422–2429, ISSN: 2321–9653, www.ijraset.com.

[4] N. Kumar, N. N. Das, D. Gupta, K. Gupta, and J. Bindra, "Efficient automated disease diagnosis using machine learning models," *Journal of Healthcare Engineering*, vol. 2021, pp. 1–13, May 2021, doi: 10.1155/2021/9983652.

[5] I. H. Sarker, "Machine learning: Algorithms, real-world applications and research directions," *SN Computer Science*, vol. 2, no. 3, Mar. 2021, doi: 10.1007/s42979-021-00592-x.

[6] B. U. Rindhe, N. Ahire, R. Patil, S. Gagare, and M. Darade, "Heart disease prediction using machine learning," *Heart Disease*, vol. 5, no. 1, May 2021, doi: 10.48175/IJARSCT-1131.

[7] D. Shah, S. Patel, and S. K. Bharti, "Heart disease prediction using machine learning techniques," *SN Computer Science*, vol. 1, p. 345, 2020. doi: 10.1007/s42979-020-00365.

[8] D. Chicco and G. Jurman, "Machine learning can predict survival of patients with heart failure from serum creatinine and ejection fraction alone," *BMC Medical Informatics and Decision Making*, vol. 20, no. 1, Feb. 2020, doi: 10.1186/s12911-020-1023-5.

[9] S. Tuli, S. Tuli, R. Tuli, and S. S. Gill, "Predicting the growth and trend of COVID-19 pandemic using machine learning and cloud computing," *Internet of Things*, vol. 11, p. 100222, 2020. doi: 10.1016/j.iot.2020.100222.

[10] N. Sharma, R. Sharma, and N. Jindal, "Machine learning and deep learning applications-a vision," *Global Transitions Proceedings*, vol. 2, no. 1, pp. 24–28, Jun. 2021, doi: 10.1016/j.gltp.2021.01.004.

[11] A. X. Du, S. Emam, and R. Gniadecki, "Review of machine learning in predicting dermatological outcomes," *Frontiers in Medicine*, vol. 7, Article 266, Jun. 2020.

[12] S. Kazeminia, C. Baur, A. Kuijper, B. van Ginneken, N. Navab, S. Albarqouni, and A. Mukhopadhyay, "GANs for medical image analysis," *Artificial Intelligence in Medicine*, vol. 109, p. 101938, 2020.

[13] D. E. Salhi, A. Tari, and T. Kechadi, "Using machine learning for heart disease prediction," in *Lecture Notes in Networks and Systems*, pp. 70–81, 2021, doi: 10.1007/978-3-030-69418-0_7.

[14] L. Yahaya, N. D. Oye, and E. J. Garba, "A comprehensive review on heart disease prediction using data mining and machine learning techniques," *American Journal of Artificial Intelligence*, vol. 4, no. 1, p. 20, Jan. 2020, doi: 10.11648/j.ajai.20200401.12.

[15] N. Ravindhar, H. S. Anand, and G. W. Ragavendran, "Intelligent diagnosis of cardiac disease prediction using machine learning," *International Journal of Innovative Technology and Exploring Engineering*, vol. 8, no. 11, Sep. 2019, doi: 10.35940/ijitee.J9765.0981119.

[16] S. J. Al'Aref et al., "Clinical applications of machine learning in cardiovascular disease and its relevance to cardiac imaging," *European Heart Journal*, vol. 40, no. 24, pp. 1975–1986, Jul. 2018, doi: 10.1093/eurheartj/ehy404.

[17] Ö. Çelik, "A research on machine learning methods and its applications," *Journal of Educational Technology and Online Learning*, vol. 1, no. 3, pp. 25–40, 2018.

[18] U. Haq, J. P. Li, M. H. Memon, S. Nazir, and S. Ruinan, "A hybrid intelligent system framework for the prediction of heart disease using machine learning algorithms," *Mobile Information Systems*, vol. 2018, pp. 1–21, Dec. 2018, doi: 10.1155/2018/3860146.

[19] F. Beritelli, G. Capizzi, G. Lo Sciuto, C. Napoli, and F. Scaglione, "Automatic heart activity diagnosis based on Gram polynomials and probabilistic neural networks," *Biomedical Engineering Letters*, vol. 8, no. 1, pp. 77–85, Aug. 2017, doi: 10.1007/s13534-017-0046-z.

[20] K. Shailaja, B. Seetharamulu, and M. A. Jabbar, "Machine learning in healthcare: A review," *Second International Conference on Electronics, Communication and Aerospace Technology*, pp. 910–914, Mar. 2018. IEEE.

[21] T. Sharma and S. Verma, "Prediction of heart disease using Cleveland dataset: A machine learning approach," *International Journal of Recent Research Aspects*, vol. 4, no. 3, pp. 17–21, Sept. 2017, ISSN: 2349-7688.

[22] A. Dey, "Machine learning algorithms: A review," *International Journal of Computer Science and Information Technologies*, vol. 7, no. 3, pp. 1174–1179, 2016.

[23] A. Simon, M. S. Deo, S. Venkatesan, and D. R. Ramesh Babu, "An overview of machine learning and its applications," *International Journal of Electrical Sciences & Engineering (IJESE)*, vol. 1, no. 1, pp. 22–24, 2015.

[24] Y. Zhang, R. Fogoros, J. Thompson, B. H. Kenknight, M. J. Pederson, A. Patangay, and S. T. Mazar, U.S. Patent No. 8,014,863, 2011. Washington, DC: U.S. Patent and Trademark Office.

[25] R. Katarya and P. Srinivas, "Predicting heart disease at early stages using machine learning: A survey," *2020 International Conference on Electronics and Sustainable Communication Systems (ICESC)*, Jul. 2020, doi: 10.1109/icesc48915.2020.9155586.

[26] R. Venkatesh, C. Balasubramanian, and M. Kaliappan, "Development of big data predictive analytics model for disease prediction using machine learning technique," *Journal of Medical Systems*, vol. 43, no. 8, Jul. 2019, doi: 10.1007/s10916-019-1398-y.

[27] S. S. Sarmah, "An efficient IOT-Based patient monitoring and heart disease prediction system using deep learning modified neural network," *IEEE Access*, vol. 8, pp. 135784–135797, Jan. 2020, doi: 10.1109/access.2020.3007561.

[28] R. Bharti, A. Khamparia, M. Shabaz, G. Dhiman, S. Pande, and P. Singh, "Prediction of heart disease using a combination of machine learning and deep learning," *Computational Intelligence and Neuroscience*, vol. 2021, pp. 1–11, Jul. 2021, doi: 10.1155/2021/8387680.

[29] V. Harikrishnan, M. Vijarania, and A. Gambhir, "Diabetic retinopathy identification using autoML," pp. 175–188, Elsevier eBooks, 2020, doi: 10.1016/b978-0-12-820604-1.00012-1.

[30] M. Rastogi, M. Vijarania, and N. Goel, "Implementation of machine learning techniques in breast cancer detection," in *International Conference on Innovative Computing and Communication*, pp. 111–121, Feb. 2023. Singapore: Springer Nature Singapore.

[31] V. K. Harikrishnan, Meenu, and A. Gambhir, "Neural AutoML with convolutional networks for diabetic retinopathy diagnosis," in *Machine Intelligence and Smart Systems, Proceedings of MISS 2020*, pp. 145–157, 2021. Singapore: Springer Singapore.

[32] R. Katarya and S. K. Meena, "Machine learning techniques for heart disease prediction: A comparative study and analysis," *Health and Technology*, vol. 11, pp. 87–97, 2021.

[33] M. Vijarania, M. Udbhav, S. Gupta, R. Kumar, and A. Agarwal, "Global cost of living in different geographical areas using the concept of NLP," in *Handbook of Research on Applications of AI, Digital Twin, and Internet of Things for Sustainable Development*, pp. 419–436, 2023. IGI Global. doi: 10.4018/978-1-6684-6821-0.ch024.

[34] S. Gupta, M. Vijarania, and M. Udbhav, "A Machine learning approach for predicting price of used cars and power demand forecasting to conserve non-renewable energy sources," in *Renewable Energy Optimization, Planning and Control: Proceedings of ICRTE 2022*, pp. 301–310, 2023. Singapore: Springer Nature Singapore.

12 Analyzing the Performance of ML Classification Algorithms for Stroke Prediction

Harshita Sharma, Richa Verma, Sunidhi Gulati

12.1 INTRODUCTION

This study provides an integrated method to forecast the possibility of a stroke. We were able to raise the prediction accuracy of our model by combining machine learning and statistical methods. It is now possible to anticipate a stroke using machine learning as a result of developments in medical science.

The primary objective of the study was to systematically review studies in each of the four categories of ML techniques for stroke based on their functionalities or similarities [1]. The results and accuracy of various machine learning models applied to text- and image-based datasets are further discussed in the study. Numerous issues related to stroke were discussed by the authors of this study.

The reviewed studies were categorized according to their similarities. The studies were hard to compare because they use different performance metrics for different tasks, look at different datasets, use different techniques, and use different tuning parameters for different tasks. Consequently, this study only mentions the research areas that were the focus of multiple studies and the studies with the highest stroke classification accuracy. When compared to a single stroke, the dataset was highly unbalanced [2]. A new study says that, using pattern recognition algorithms, machine learning (ML) is becoming an important tool for diagnosing, treating, and predicting complications and patient outcomes in several neurological diseases. In the final section of the chapter, we talk about how machine learning applications are flourishing in the AIS field, which is expanding rapidly and becoming increasingly dependent on neuroimaging. ML solutions are of particular importance in this field, where the combination of intricate data and a scarcity of human experts presents a significant challenge. Future ML for AIS headings may require cooperation between various foundations to create a strong dataset for productive ML network preparing. A semantic analysis system for early stroke detection and recurrence in people over 65 was developed by the researchers with the help of the NIH stroke scale. The research was made possible by the data that were gathered from Kaggle, which included 5,110 stroke patients.

DOI: 10.1201/9781003477280-12

The proposed framework partitions the stroke seriousness score into four classes for arrangement and forecast utilizing agent AI and information mining strategies. In tests, the precision of the proposed framework is estimated utilizing review and accuracy estimations [3]. Thanks to a variety of machine learning algorithms, the experiment produced faster, more precise stroke severity predictions and system operation efficiency. The chapter's decision was that it was clear how AI systems and a reasonable dataset can be utilized to construct a model to anticipate strokes and to further evaluate the seriousness of side effects to anticipate results with high accuracy and create a system that provides real-time alerts advising people to visit a clinic or hospital. The dataset, which had 12 columns and 5,110 rows, attributes the terms "stroke" attribute due to "0" (no stroke) and "id," "gender," "age," "hypertension," "heart disease," "ever married," "work type," "residence type," "avg_ glucose level," "BMI," "smoking status" were used to train the model for stroke prediction using KNN, random forest, logistic regression, and decision trees [4]. The accuracy achieved by the three approaches was comparable to that of ML models that call for both outcomes (labels) and predictors. Learning without supervision is a subset of machine learning models that use predictors or variables to identify previously unidentified patterns in data without using labels.

Feature selection: The ability to select variables or attributes refers to the capacity to apply a previously learned pattern to new data.

The instruction: The method by which a model learns the data pattern.

Testing: A set of validations that are put to the model's test. Support vector machine, also known as SVM, is a supervised classifier that looks for the best hyperplane for data separation. KNN, or k-nearest neighbours, is a type of instance-based learning in which only the closest neighbours locally are used to make predictions. An artificial neural network (ANN) is a type of computational model that is built on a collection of connected units or nodes. The neurons that are found in a living brain are loosely represented by these artificial neurons.

Decision tree: A collection of tree-based decisions that eventually lead to a decision.

Random forest: A method of ensemble learning that makes use of numerous decision trees is called random forest.

12.2 LITERATURE SURVEY

There are two primary drivers of stroke: a draining or blasting vein (hemorrhagic stroke) or an obstructed course (ischemic stroke). In some people, a transient ischemic attack (TIA), also known as a brief interruption in blood flow to the brain that does not result in a stroke that is permanent, may occur [5].

A stroke is characterized as an intense cerebral, spinal, or retinal vascular mishap with neurological brokenness enduring more than 24 hours while imaging (figured tomography attractive reverberation sweeps) or dissection exhibits localized necrosis or related side effects. According to a recent survey conducted by the World Health Organization (WHO), cardiovascular diseases (CVDs) account for approximately

17.9 million deaths globally on an annual basis [6]. Four out of every five CVD deaths are caused by heart attacks and strokes, and this number is rapidly rising. When blood vessels in the brain break or become blocked, this results in a stroke. A precise expectation can be useful for both early stroke risk recognition and treatment.

Stroke is the second leading cause of death worldwide. Strokes kill five million people annually, according to the World Health Organization. One-third of them die, and another third are left with disabilities that last a lifetime. A stroke occurs every 40 seconds, and death occurs every four minutes [7].

As a result of a disruption in the flow of blood to the brain, victims experience a wide range of incapacitating symptoms, including sudden paralysis, speech loss, and blindness [8]. A stroke can cause incapacities that are either brief or extremely durable, depending on how long the cerebrum is without blood flow and what part is impacted. Realizing your stroke risk factors and heeding your primary care physician's guidance are solid protection measures. If you have already suffered a heart attack or a transient ischemic assault (TIA), these actions might help prevent it from recurring.

12.3 PROPOSED METHODOLOGY

Machine learning is becoming increasingly popular in the field of stroke medicine. Even though there are numerous machine learning algorithms, selecting the best one for stroke disease datasets, such as the proposed K-nearest neighbors or support vector classification, is still challenging.

Model building requires a preprocessed dataset and machine learning algorithms. The steps are as follows:

12.3.1 Dataset

The Kaggle dataset is used to foresee whether a patient will have a stroke by taking into account subordinate factors like orientation, age, an assortment of medical issues, and smoking status. The remaining eight features were categorical, consisting of three numerical and continuous features, seven dependent features, and one independent feature. Relevant information about the patient can be found in each row of the data. Because "stroke" was a binary feature, classification modeling was used. We chose participants over 18 from the Kaggle dataset for our study [9]. The following descriptions are provided for each attribute (ten as input to ML models and one for the target class) of the 3,254 participants:

1. Age in years: This feature focuses on people over the age of 18 who participate in the study.
2. Gender: This feature addresses the gender of the participant. The population includes 1,260 men and 1,994 women.
3. Tension in the body: This trait can indicate whether a subject is hypertensive; 12.54% of participants have high blood pressure.
4. Work type: There are four categories for this quality: private (65.02%), self-employed (19.21%), public (15.67%), and never worked (0.1%).

5. Dwelling: This feature is broken down into two categories: urban (51.14%) and rural (48.86%), representing the participant's living situation.
6. Average glucose in the blood: The participant's average blood glucose level is recorded by this feature.
7. BMI (kilos/m2): The participant's BMI is recorded by this feature.

12.3.2 Data Preprocessing

Data preprocessing is an essential step in accurately preparing the data for the machine learning algorithm [10]. It is essential for enhancing the performance outcomes of machine learning. Several steps are taken at this stage, beginning with an examination of smoking status and features with numerous missing values. The mean is used to fill in the missing values before Label Encoder turns class labels into quantitative values. This database's data is skewed. The basic methodology for data preprocessing is shown in Figure 12.1. An unbalanced ratio of values for each class label indicates that the data is unbalanced. Random resampling methods are used to deal with data with imbalances [11].

There are a lot of noisy and missing values in the actual data. In order to avoid such issues and make accurate predictions, these data are preprocessed: i.e. the raw data is inconsistent and insufficient, so the mean value can be used to fill in missing values. In this manner, to direct an effective examination, the information obtained should be marginally changed utilizing a separating technique. For this, a multi-shifting technique is utilized.

12.3.2.1 Label Encoding

The textual expressions in the dataset are transformed by label encoding into machine-readable constant values. The strings must be encoded into integers because the computer is typically trained with numbers. Strings are contained in five columns in the compiled dataset. All the strings are encoded when label encoding is used, transforming the entire dataset into a series of numbers, as shown in Figure 12.1.

12.3.2.2 Libraries Imported

This project made use of a number of libraries, including Numpy, Pandas, Seaborn, Matplotlib, and Sklearn. We imported random forest classifier, SVM, KNN, F1 score, accuracy score, classification report, and precision score from Sklearn as well.

12.3.2.2.1 Random Forest Classification

The classification algorithm employed was RF classification. In a random collection of data, numerous decision trees were separately trained to create RFs. These trees are constructed during training, and classification tree outcomes are retrieved. The

FIGURE 12.1 Basic methodology.

algorithm's final prediction is made through a process called voting. Every DT in this approach must select between the two discrete classes (in this scenario, stroke or no stroke), and the RF method chooses the class with the most votes to generate the final prediction.

The versatility of the random forest is among its most appealing aspects. The weighting of the data features is observable, and it may be applied to applications like categorization and relapse detection. It is beneficial since this typically employs default hyperparameters that provide specific expectations. It's critical to be aware of them because there are initially a limited number of them. Overfitting is a well-known difficulty in machine learning, yet the arbitrary random forest classifier seldom runs into it. By ensuring an adequate number of trees in the forest, the classifier can mitigate the risk of overfitting the model.

12.3.2.2.2 K-Nearest Neighbors Classification

The K-nearest neighbors (KNN) classifier uses vicinity to make predictions or label distinct data points. The results of this method will vary for both classification and regression scenarios. In cases of classification, a decision is made by majority vote, which means the new data point's class is taken to be the one that is seen most frequently in its vicinity.

```
Random Forest Classification's Accuracy: 0.9547325102880658
K-Fold Validation Mean Accuracy: 95.55 %
Standard Deviation: 0.20 %
              precision    recall  f1-score   support

           0       0.96      1.00      0.98       929
           1       0.00      0.00      0.00        43

    accuracy                           0.95       972
   macro avg       0.48      0.50      0.49       972
weighted avg       0.91      0.95      0.93       972
```

FIGURE 12.2 Random forest classification results

```
K-Nearest Neighbours's Accuracy : 0.9557613168724279
K-Fold Validation Mean Accuracy: 95.75 %
Standard Deviation: 0.12 %
              precision    recall  f1-score   support

           0       0.96      1.00      0.98       929
           1       0.00      0.00      0.00        43

    accuracy                           0.96       972
   macro avg       0.48      0.50      0.49       972
weighted avg       0.91      0.96      0.93       972
```

FIGURE 12.3 K-NN results.

```
Kernel SVM's Accuracy : 0.9557613168724279
K-Fold Validation Mean Accuracy: 95.75 %
Standard Deviation: 0.12 %
              precision    recall  f1-score   support

           0       0.96      1.00      0.98       929
           1       0.00      0.00      0.00        43

    accuracy                           0.96       972
   macro avg       0.48      0.50      0.49       972
weighted avg       0.91      0.96      0.93       972
```

FIGURE 12.4 SVM results.

12.3.2.2.3 Support Vector Machine (SVM)

One of the really prominent classification models is the support vector machine or SVM. Regression and classification problems can be solved with this. However, classification issues in machine learning are its primary application. The objective of the SVM calculation is to decide the proper choice limit or line for arranging the n-layered space, enabling us to swiftly classify a fresh data point in the future. The terminology for this optimum decision boundary is a hyperplane. SVM chooses the extreme ends and vectors that help in the formation of the hyperplane. It is a technique for supervised learning that can be combined with learning algorithms for regression and data classification. When scaled up, the SVM manages high-dimensional data reasonably well.

A confusion matrix is the simplest method for comparing the number of instances that were correctly or incorrectly.

F1 score: The F1 score for accuracy testing is the harmonic mean of recall and precision.

Using randomly generated five-fold datasets, the entire procedure was repeated 20 times to compare the ML models' performance.

We utilized the permutation method for the testing dataset's random forest model to estimate each predictor's significance. We used the permutation method to find important features because changing their values in the dataset would cause the model to make more wrong predictions. The machine learning model predicts the output once it has the input parameters. The flask application then reflects the result back onto the web page, allowing the user to view the prediction.

12.4 RESULTS AND DISCUSSION

Medical data analysis is very concerned with stroke disease performance because it is one of the most prevalent causes of death. AI can possibly work on specialists' bits of knowledge, especially in the forecast of coronary illness, permitting them to better adjust patient analysis and treatment.

This chapter looks into the viability and value of various machine learning algorithms. Forecasting cardiovascular stroke is among the main concerns in clinical

TABLE 12.1
Optimal Stroke Accuracy Evaluation

Machine Learning	Optimal Stroke Accuracy	AUC (Stroke/Healthy)	Accuracy (Gait Pattern)
Support Vector Machine	**95.57%**	**0.999**	**95.57%**
K-Nearest Neighbors	**95.05%**	**0.700**	**95.75%**
Random Forest	**95.4%**	**0.952**	**95.47%**

data analysis, considering that heart disease is currently one of the top causes of death. The outcomes of strokes are influenced by a variety of factors, and these variables may even have a small impact on prediction.

Strokes seemed more prevalent in areas with a higher percentage of non-Hispanic Black residents, while stroke prevalence was lower in rich areas. Additionally, we noticed that an older demographic composition and a lack of recreational physical exercise had a synergistic impact on the prevalence of stroke at the local scale. Community-level initiatives that promote exercise, enhance air quality, and create exercise-friendly neighborhoods may lower stroke rates. The SVM model's good predictive ability, as demonstrated by our findings, will be of clinical use in determining the risk of SVE six months after MIS.

It might be best to hold machine learning tools to the same standards as any new medical diagnostic technology, like a new imaging or laboratory test. The classification algorithms' computational results ought to be compared to gold standard measures, and their predictions ought to be validated using long-term, high-quality follow-up data. Five performance parameters and the values of the confusion matrix are listed; two classifiers are boldly highlighted.

The SVM model had an optimal stroke accuracy of 95.57 percent when all model parameters were taken into account. From Figures 12.5 and 12.6, SVM's calculated accuracy was 95.57%, KNN's was 95.05%, and random forest was 95.4% percent.

The area under the curve (AUC) of various machine learning algorithms for classifying stroke patients' and healthy normal subjects' gait patterns shows the highest classification accuracy at 0.999 0.002; KNN, SVM, and random forest show accuracy of 95.75%, 95.57%, and 95.47%, respectively.

12.5 CONCLUSION

Before it intensifies, a stroke is a hazardous medical condition that has to be addressed. The creation of a machine learning model may assist with early stroke diagnosis and decrease its severe future effects. Based on a wide range of physiological parameters, this chapter demonstrates that several machine learning algorithms can accurately predict stroke. With an accuracy of 95.75%, SVM outperforms the other algorithms chosen.

ML is increasingly being used to predict the outcomes of strokes. However, none of them met the fundamental reporting requirements for clinical prediction tools, nor

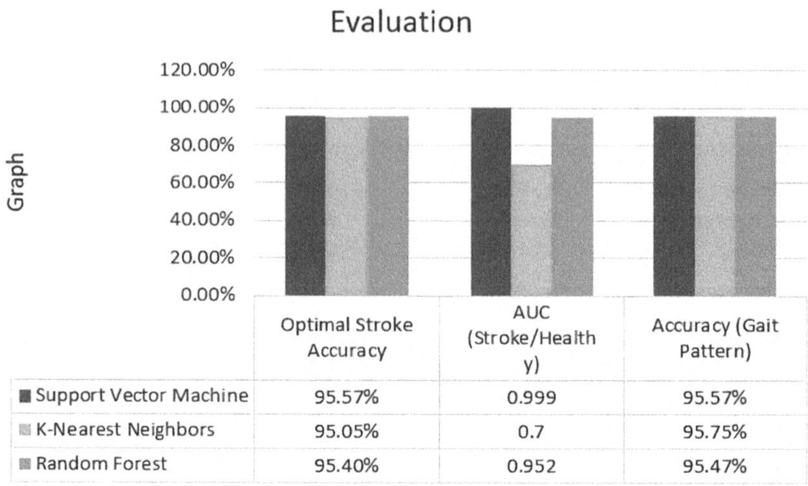

FIGURE 12.5 Bar representation of evaluation report.

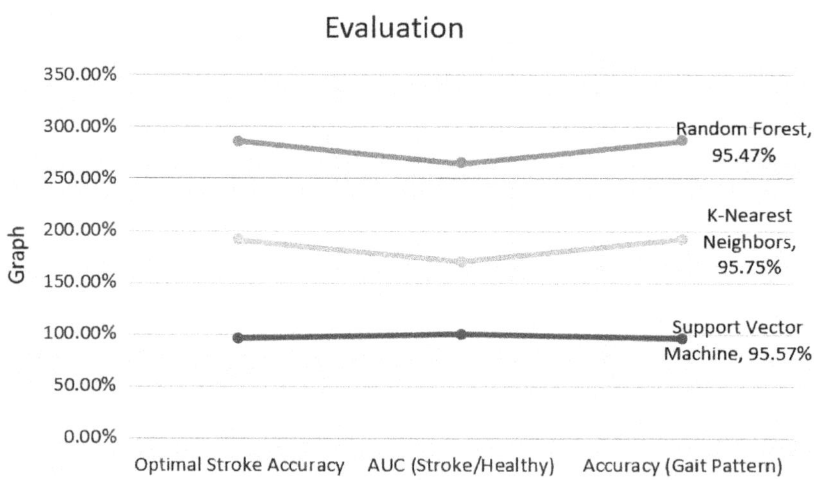

FIGURE 12.6 Line representation of evaluation report.

did they make their models available in a way that could be used or evaluated. Before it can be meaningfully considered for use in practice, significant improvements in the conduct and reporting of ML studies are required. Age, past ED visits over a year, pre-stroke useful status, beginning stroke seriousness, BMI, cognizance level, and utilization of nasogastric tubes were undeniably observed to be critical indicators of readmission or mortality. Moreover, Figure 12.5 and Figure 12.6 show us the clear trend of how strokes are related to different attributes.

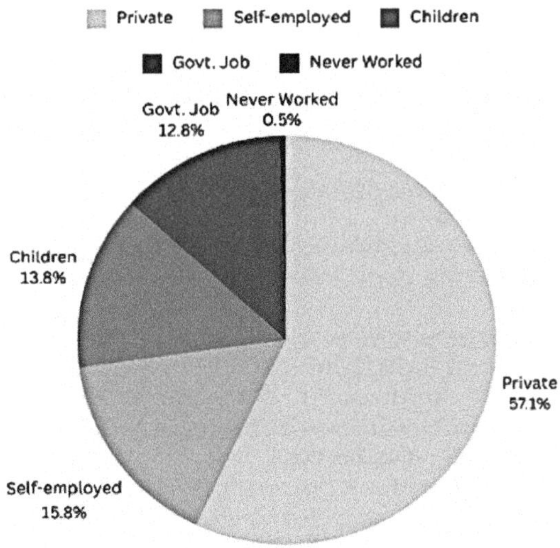

FIGURE 12.7 Stroke trend on the basis of work type.

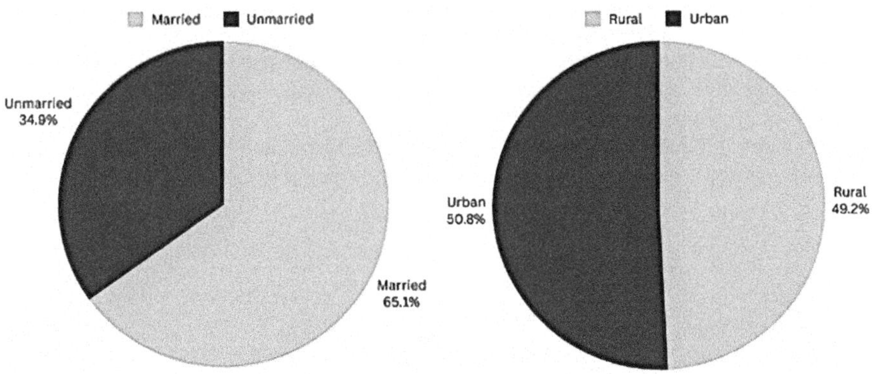

FIGURE 12.8 Stroke trend in married vs unmarried and urban vs rural subjects.

The availability of large, high-quality datasets is essential to machine learning. Due to the increasing availability of large data, health policy is a particularly appealing field.

Predictive algorithms, on the other hand, run the risk of doing less good than they otherwise might if we do not take the measurement process and generation process seriously. They may even cause harm in some instances, and machine learning techniques are able to accurately quantify proximal arm weakness and identify signs of stroke.

REFERENCES

[1] Singh, M. S., Choudhary, P., Thongam, K. A comparative analysis for various stroke prediction techniques. In *Computer Vision and Image Processing*. Springer, 2020.

[2] Pradeepa, S., Manjula, K. R., Vimal, S., Khan, M. S., Chilamkurti, N., Luhach, A. Kr. DRFS: Detecting risk factor of stroke disease from social media using machine learning techniques. *Neural Processing Letters*, 55:3843–3861, 2023. doi: 10.1007/s11063-020-10279-8. ISSN: 1370-4621.

[3] Bandi, V., Bhattacharyya, D., Midhunchakkravarthy, D. Prediction of brain stroke severity using machine learning. *Revue d'Intelligence Artificielle*, 34(6):753–761, 2020. doi: 10.18280/ria.340609.

[4] Tibshirani, R. Regression shrinkage and selection via the lasso. *Journal of the Royal Statistical Society, Series B*, 58(1):267–288, 1996.

[5] Vokó, Z., Hollander, M., Koudstaal, P. J., Hofman, A., Breteler, M. M. How do American stroke risk functions perform in a western European population? *Neuro Epidemiology*, 23(5):247–253, September–October 2004.

[6] Nwosu, C. S., Dev, S., Bhardwaj, P., Veeravalli, B., John, D. Predicting stroke from electronic health records. In *41st Annual International Conference of the IEEE Engineering in Medicine and Biology Society*. IEEE, 2019.

[7] Ng, Y. Feature selection, l1 vs. l2 regularisation, and rotational invariance. In *Proceedings of the International Conference on Machine Learning*, 2004.

[8] Park, M.-Y., Hastie, T. An l1 regularisation-path algorithm for generalised linear models. *Journal of the Royal Statistical Society: Series B*, 69(4):659–677, 2007.

[9] Raykar, V., Steck, H., Krishnapuram, B., Dehing-Oberije, C., Lambin, P. On ranking in survival analysis: Bounds on the concordance index. In *Advances in Neural Information Processing Systems*, Vol. 20. MIT Press, 2008.

[10] Schmidt, M., Fung, G., Rosales, R. Fast optimization methods for l1 regularisation: A comparative study and two new approaches. In *Proceedings of the European Conference on Machine Learning*, 2007.

[11] Schneider, T. Analysis of incomplete climate data: Estimation of mean values and covariance matrices and imputation of missing values. *Journal of Climate*, 14:853–871, 2001.

Index

Note: Page numbers in **bold** indicate a table and page numbers in *italics* indicate a figure on the corresponding page.

For Product Safety Concerns and Information please contact our EU
representative GPSR@taylorandfrancis.com
Taylor & Francis Verlag GmbH, Kaufingerstraße 24, 80331 München, Germany

www.ingramcontent.com/pod-product-compliance
Ingram Content Group UK Ltd.
Pitfield, Milton Keynes, MK11 3LW, UK
UKHW022123190926
475702UK00013B/1613
* 9 7 8 1 0 3 2 7 6 1 4 7 3 *